WOODLOT
MANAGEMENT

WOODLOT MANAGEMENT

BRUNO WISKEL

First printed in 1995 5 4 3 2 1

Printed in Canada

The Publisher

Lone Pine Publishing

#206, 10426-81 Avenue
Edmonton, Alberta, Canada
T6E 1X5

202A, 1110 Seymour Street
Vancouver, British Columbia, Canada
V6B 3N3

16149 Redmond Way, #180
Redmond, Washington, USA
98052

Canadian Cataloguing in Publication Data

Wiskel, Bruno.
Woodlot management

ISBN 1-55105-067-6
1. Woodlots–Canada–Management. 2. Woodlots–Management. I. Title.
SD387.W6W57 1995 634.9'9 C95-911039-9

Editorial: *Nancy Foulds*
Design and layout: *Bruce Timothy Keith, Greg Brown*
Cover design: *Bruce Timothy Keith*
Cover background photograph: *Dean Dragich*
Printing: *Best Book Manufacturers*

Dedication

This book is dedicated to the memory of

S.N. HORNER
Creelman, Saskatchewan

"YOU ARE GONE, BUT
YOUR FOREST LIVES ON."

CONTENTS

ACKNOWLEDGEMENTS

Sincere appreciation is given to Alberta-Pacific Forest Industries Inc. (Al-Pac), especially Doug Sklar, for providing financial and technical support without which this book could not have been possible. Special thanks to Dan MacPherson of Al-Pac and Jim Pearson of Pearson Timberline for the many hours of technical consultation spent in developing this book, and Al-Pac's Pam Strausz for providing the numerous illustrations. I am also grateful to Bob Raina, Tim Kemp and Mark Leriger of Northern Envirosearch for their financial assistance.

Special thanks also to my brother Barry for his enormous contribution to the reforestation section; to my father Stan for letting me use his tractor for harvesting when mine would not start in the winter; to my mother Sandy for providing moral support and still not being ashamed to admit I am her son; and to my brother Bruce and his family who always seem to help me out no matter what I am doing. Last but not least, special thanks to Lasha Turner for the many hours of assistance in completing the final draft.

Others who provided technical expertise are listed in no particular order: Byron Grundberg and John Kort at Prairie Farm Rehabilitation Administration for supplying valuable provincial information and for facts about shelterbelts, Grant Williamson at the Canadian Forest Service, Mike Carlson at the Kalamalka Lake Forest Research Station for material on urban forestry, Mike Gillis of New Grazing Partnership Ltd. for details on sheep grazing, John Froe (park naturalist at Moose Mountain Provincial Park) for bird-watching instruction, and Bill Stilwell and Patricia E. Mackonka for woodcarving and jewellery-box building demonstrations.

A whole whack of people had only minor roles in the book's production but still wanted to see their name in print: Jim Kostyck, Dooley Nelson, Ruby Backman, David Huff, Tim Holden, Walter and Kay Tostiniuk, Moe and Jo Breckenridge, Roman Bizon, Murray Lyle, Norm and Bev Gayley, Ed Ramsum, Dennis Stady, Peter Behrens, Eli Pederson, Randy and Paula de Bruijn and family, Alan Fisher, Rod Kowalchuck, Alan Salzl, Terry Francis, Florian Skuban and Deanna McCullough.

Thanx to you all.

INTRODUCTION

Close your eyes and imagine some of the most joyous times in your life: a child's swan dive into a pile of freshly raked leaves; a romantic walk for two down a tree-lined lane; waking to the sound of leaves rustling and birds chirping on a summer camping trip; or enjoying a cup of hot cocoa in front of a crackling fire after a brisk mid-winter ski. All of these images of peace and happiness are connected by one entity, the forest.

Now create a mental picture of your ideal lifestyle. Perhaps it includes a new house, or a job that is both mentally stimulating and physically invigorating. Perhaps it is the satifaction of knowing that by your efforts the world is a little better place to live. The forest can make all these dreams a reality.

Woodlot Management is a complete guide to sustainable small-scale forest management, the practice of which will provide endless hours of memorable recreation for the entire family *and* furnish a primary or supplementary source of income. This book explains how forests and industry can be compatible and can mutually benefit each other.

Woodlot Management is for people who own a woodlot that contains one tree or more. It gives the woodlot owner realistic options, uses recognized forestry principles and proposes some new radical alternatives to forest management which are applicable to both urban and rural dwellers.

The woodlot where most of the research for the book was completed lies in the Little Pine Creek valley just east of Colinton, Alberta. This woodlot has provided loads of lumber, cords of firewood and a bounty of berries, mushrooms and syrup, yet the forest animals continue to shelter there each winter and the songbirds return each spring. This woodlot has provided thousands of dollars of additional revenue and, more importantly, has provided countless hours of natural entertainment, walking, camping, skiing and sitting watching the beavers and the birds going about their business.

If, by reading this book, you do nothing more than begin to realize that a woodlot, even a very tiny woodlot, can be profitable both financially and spiritually, then all the work that has been put into this book has been worthwhile. Put up the hammock and enjoy!

Chapter One

A HISTORY OF FORESTS IN CANADA

Canada is widely regarded at home and abroad as a forest nation. The forest and its inhabitants are represented around the world on our flag and our currency. The painters have painted it, the singers have sung about it, and industry has prospered from it. The Canadian forest, in fact, is a big part of what underlies our national identity. With such emphasis on the forest as part of the national consciousness, it may come as a surprise to learn that the land we now call Canada has been without its green cloak of forest many times in the past.

At the time this book was written, Canada was well into its 128th year, so it may be a revelation to some to learn that the "modern" forest had its humble beginnings some 12,000 years ago with the retreat of the glaciers. The first forest in what is now Canada began more than 400 million years before that!

To understand the concept of time, trees and the development of forests in Canada, it is helpful to use the metaphor represented in the 24-hour day. If it is midnight right now, the first land plant grew in Quebec at 12:00 AM the previous day. The lush fern, ginkgo and cycad forests that provided material for the New Brunswick and Nova Scotia coal fields grew at 5:00 AM. Sub-tropical forests containing such recognizable species as pine, birch and willow covered large areas of B.C., Alberta and Saskatchewan and later formed coal beds; these forests grew from about 5:30 to 7:30 PM. (7:30 PM was also bedtime for the dinosaurs, a sleep from which they never awoke.) The first modern Canadian forest appeared less than three seconds ago and commercial logging less than 1/10 of a second before midnight.

24-Hour Clock	Canadian Forest	Years Before Present
12:00 AM	First land plant grows in Quebec	400 million
1:00 AM		
2:00 AM		
3:00 AM		350 million
4:00 AM		
5:00 AM	New Brunswick/Nova Scotia fern, cycad & ginkgo forest now present as coal beds	320 million
6:00 AM		300 million
7:00 AM		
8:00 AM		
9:00 AM		
10:00 AM		
11:00 AM		
12:00 Noon		200 million
1:00 PM		
2:00 PM		
3:00 PM		
4:00 PM		
5:00 PM		
5:30 PM	Alberta/Saskatchewan/B.C. pine, birch & willow forest now present as coal beds	110 million
6.00 PM		
7:00 PM		
7:30 PM	Extinction of the dinosaurs	66 million

Log cabins were home to many of Canada's early immigrants. The walls, roofs and much of the furniture came directly from the surrounding forest. (Courtesy Lake of the Woods Museum.)

8:00 PM	
9:00 PM	
10:00 PM	
11:00 PM	
12:00 PM	Glaciers recede at 11:59:57. 3 seconds = 12,000 years

MODERN PRE-HISTORICAL CANADIAN FORESTS

24-Hour Clock	Forest	Years Before Present
12:00 AM	Receding of the glaciers	12,000
1:00 AM	Northward expansion of the forest	
2:00 AM		11,000
3:00 AM		
4:00 AM	First aboriginals enter Canada	10,000
5:00 AM		
6:00 AM	Aboriginals use forests	9,000
	for all their needs: fuel, shelter, hunting grounds	
7:00 AM		
8:00 AM		8,000
9:00 AM	The Haida construct totem poles	
10:00 AM		7,000
11:00 AM		
12:00 Noon		6,000
1:00 PM		
2:00 PM		5,000
3:00 PM		
4:00 PM		4,000
5:00 PM		
6:00 PM		3,000
7:00 PM		
8:00 PM	Christ was born	2,000

From the landing, logs were floated downstream and held in booms outside the mills before processing. (Courtesy Lake of the Woods Museum.)

9:00 PM		
10:00 PM		1,000
11:00 PM	Modern Historical Canadian Forests (expanded) 24 hours equals 500 years	500
12:00 Midnight		Present Day

MODERN HISTORICAL CANADIAN FORESTS

24-Hour Clock	Forest	Year
	Columbus reaches America	1492 AD
12:00 AM		1500
1:00 AM	Jacques Cartier reaches Canada	1534
2:00 AM		
3:00 AM		
4:00 AM		
5:00 AM	Samuel de Champlain establishes settlement at mouth of St. Croix River	1604
6:00 AM	Small sawmills operational along St. Lawrence River to supply lumber for domestic markets	
7:00 AM		
8:00 AM	First export of Canadian timber to France	1667
9:00 AM		
10:00 AM	Timber exports dwindle to nothing because of poor Canadian sawing techniques	1700
11:00 AM		
12:00 Noon		

Except for the dinosaurs, ancient forests often did not look a whole lot different than forests today.

1:00 PM		
2:00 PM	First timber exports to British Navy	1779
	Napoleon's Baltic Blockade expands British imports 100 times	1799
3:00 PM	Ottawa Valley opened for logging white pine	1800
4:00 PM	First railway between La Prairie and St. Johns, Quebec	1832
5:00 PM	Start of B.C. lumber industry to supply California gold rush	1850
	Start of massive railway building in Canada	1852
	First attempts to establish forest fire control	1854
6:00 PM	First pulpmill opened in Valleyfield, Quebec	1866
	Lumber grading system established	1870
7:00 PM	Canadian Pacific Railway started	1881
	Algonquin Park, Ontario, first multiple use forest	1893
	Forest Reserve Act protecting Alberta east slopes	1898
	Canadian Forest Service inaugurated	1899
8:00 PM	First professional forester hired, Judson Clarke	1902
	First forestry school opened at University of Toronto	1907
	U.S. replaces Britain as #1 importer	1908
9:00 PM	First Royal Commission study on woodlands	1924
	Depression causes major mill, reforestation shutdown	1929
	Alberta, Saskatchewan & Manitoba given provincial control of their forests	1932
10:00 PM	Start of WWII increases need for wood & paper	1940
	End of WWII creates economic boom for construction material	1945
	First forest managed for sustained yield (NW Pulp & Power)	1956
11:00 PM	Forestry elevated to ministerial position	1963
	Provincial woodlot associations begin to form	1965
	Reed Report shows Canada's forests greatly diminished	1977
	Reed Report causes changes to provincial forest policy	1979
12:00 PM	Domtar pays woodlot owners to maintain forests	1982
	Spotted owl controversy restricts U.S allowable cut	1987

It's 12:00 o'clock—do you know where your forest is? 2000

Tree Trivia

Some trees living today have considerable history. The world's oldest trees are bristlecone pines found in California. Several individuals are known to be over 5000 years old!!!

Aboriginal Canadians used the forests in almost every aspect of their lives, from simple tipi support poles to ornate totem poles.

Chapter Two

ALL ABOUT TREES

Learning to manage a woodlot is a lot like learning to drive a motor vehicle. If a person is to drive effectively, it is important to understand how the vehicle's components function before taking it out on the road. An attempt to operate a vehicle without the slightest knowledge of the accelerator, brakes or rules of the road is bound to lead to a bad accident.

Likewise, to effectively manage a woodlot it is necessary to know how tree's component function before performing woodlot operations. An attempt to harvest or regenerate the forest without proper knowledge of the roots, branches and leaves may result in a bad environmental accident.

Tree species, like motor vehicle models, come in a wide variety of shapes and sizes, each with its own unique features. To simplify the purchase of a vehicle, a dealer will divide the inventory into cars or trucks. Similarly, for simplicity, the woodlot trees will be classified into only two groups: conifers (evergreens) and deciduous (broad-leaf trees).

Conifers first appeared on Earth about 300 million years ago. They are characterized by their needle-like leaves and seed-carrying cones. Timber derived from conifers is called "softwood." There are 31 species of conifers native to Canada. Families of coniferous trees found in Canada include pine, larch (tamarack), spruce, hemlock, fir, yew, cypress, juniper and cedar.

Deciduous trees came onto the scene along with the other flowering plants just about when the dinosaurs reached their peak, about 144 million years ago. Deciduous trees are characterized by summer production of leaves followed by leaf loss before winter, and a seed protected by a capsule. Timber derived from this group of trees is called "hardwood." Over 100 species of deciduous trees are native to Canada.

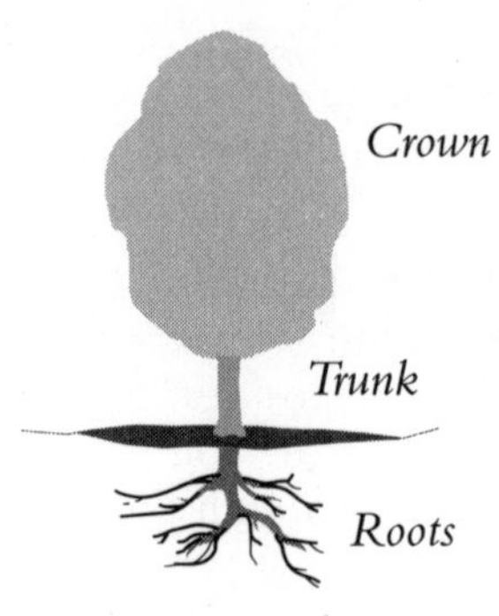

The most common families include willow, poplar, walnut, hickory, birch, alder, chestnut, oak, elm, laurel, maple, basswood, dogwood, ash and elder.

Despite differences in leaf type and seed coat, the conifers and deciduous trees have similar physiology and physical requirements. Trees, whether conifer or deciduous, can be divided into three basic components: roots, trunk and crown (SEE LEFT).

The Roots

The roots are like underground branches of the tree and are responsible for a number of functions including anchoring, nutrient uptake, storage and, in some cases, reproduction.

As the crown of the tree grows, so do the roots. At the end of each root is a tough cap that protects the rootlet as it pushes its way through the ground. Trees that grow in drier, open areas typically have roots that grow downward to stabilize the tree against the wind and to reach a reliable water supply.

Spruce trees have very shallow root systems because the deciduous overstorey shades the ground and prevents severe soil drying to occur. The shallow roots are extremely susceptible to abrasion caused by grazing cattle and motorized farm or logging equipment. The surrounding deciduous trees also provide wind protection, therefore unrestricted deciduous removal will leave the remaining spruce vulnerable to blowdown.

It is important to remember when transplanting trees that the roots will extend the same length out from the trees as the branches. Trees with a larger crown have a larger root system. To avoid large amounts of root damage, dig the hole around the tree at least the same distance from the trunk as the branches (SEE OPPPOSITE PAGE).

The roots are also important for water and nutrient uptake. The large roots visible at ground level serve the same function as the trunk: a water/nutrient pipeline. The water and nutrients are actively taken in by the myriad of root hairs at the end of the rootlets. Because of their fine texture, many root hairs are torn off or badly damaged during transplanting. Hence, the plant's ability to uptake water and dissolved minerals is greatly reduced. It is important to water trees after transplanting.

The nutrients taken up by the roots can be divided into two groups: macronutrients, which constitute over .5% of the plant's dry weight, and micronutrients, which are used by the plant in the order of parts per million.

The macronutrients other than carbon, oxygen and hydrogen are nitrogen, potassium, calcium, phosphorus, magnesium and sulphur. These nutrients are formed in the soil from the decomposition of the bedrock. Up to 40% of macronutrients are returned to the soil in the leaf and needle litter. Macronutrients are also recycled through the decay and burning of wood.

There are about 16 to 18 different micronutrients required depending on species of tree. Because micronutrients are also recycled through the fallen needles and leaves, the best way to avoid nutrient deficiency in newly planted forests is to return the land to the same species of trees that were harvested.

Monoculture (forests with only one species of tree) is about 20% less productive than a mixed-wood forest. The mixed wood has a higher production because different tree species are able use different nutrients at different rates. The mixed wood also tends to have fewer disease and insect problems because the conifers typically are not host to scourges that plague deciduous trees and vice versa.

The needle litter produced by a pure conifer forest causes the soil below to become acidic. As the soil acidity level increases, so does nutrient mobility. The mobile nutrients are susceptible to leaching during periods of heavy rainfall. Deciduous trees help neutralize the soil, thus reducing nutrient mobility and the likelihood of leaching.

Branches are a good indicator of the lateral extent of the root growth.

The roots are also places for nutrient storage. The carbohydrates produced by the tree during the growing season form a solution with water and other nutrients into sap. This sap moves through the trunk to the roots in the fall. In the spring the direction of travel is reversed as the sap moves up to provide nourishment to the newly formed leaves.

Reproduction by suckering in poplar trees.

The roots of some trees, particularly the poplars, are important for asexual reproduction. Disturbances such as fire or logging produce soil conditions conducive to the production of saplings from the roots, a process commonly called "suckering." The daughter trees produced by suckering (SEE LEFT) are genetically identical to the "mother" tree.

Poplar, a common nurse crop in the forest succession, is extremely shade intolerant. As the conifer understorey grows, light to the poplars becomes restricted, resulting in high rates of mortality among the poplar. As the poplar die off, the nutrients are returned to the roots where they can be conserved for many years. Many pure stands of conifer are actually criss-crossed with poplar roots that periodically send up suckers to "test the water." These suckers repeatedly die off and sucker again until the conifer canopy is removed and high light conditions prevail.

Tree Trivia

Suckering is so important in aspen reproduction that many northern stands several hectares in area are all genetically identical, having originally started from a single seed!!!

The Trunk

The trunk is the body of the tree. It provides strength to support the crown and contains the pipelines for food and fluid movement within the tree. These pipelines are the tree's vascular system and are shown on page 11.

The outer bark is the tree's skin. The function of the bark is to provide an unbroken membrane to keep diseases, insects and parasites on the outside, and to protect the growing tissue on the inside. The bark can be damaged in several ways, such as by fire, frost, physical abrasion and outright breakage.

Fire damage is most severe to immature trees, which often have thin and succulent bark. Wounds to the base of the tree, caused by low intensity fires, are often the entry points for cankers and other disease-

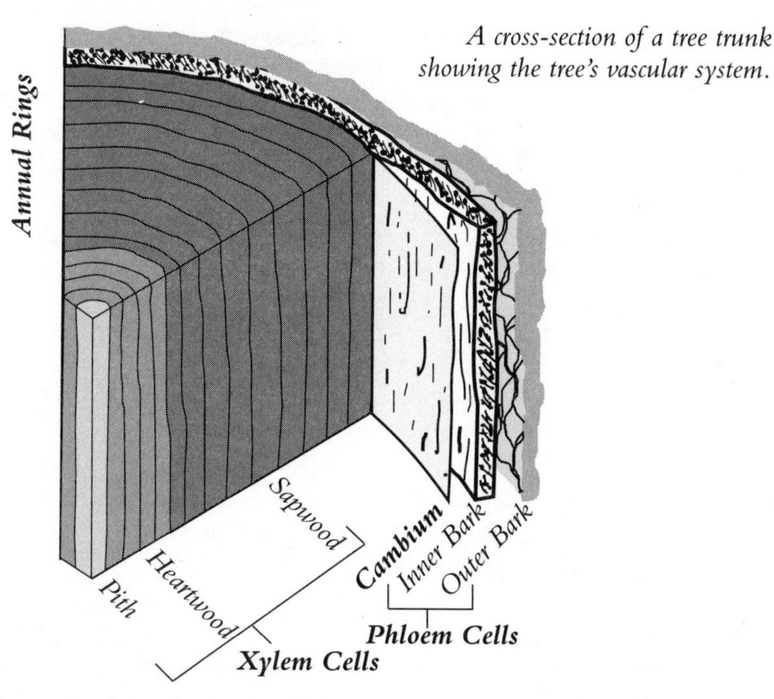

A cross-section of a tree trunk showing the tree's vascular system.

bearing organisms. Severity of the injury varies directly with heat intensity. Typically a fire will have a heat output proportional to the amount of debris on the ground.

Frost damage occurs when water crystallizes inside the living cells during rapid drops in temperature. The damaged tissue cracks as the water expands into ice.

Physical abrasion and outright breakage are most commonly caused by mechanical equipment coming into contact with the trunk, although wind action may cause rubbing between two closely spaced trees. Trees are most easily damaged by physical abrasion from the time the first bud bursts in the spring to midsummer, when the cambium (the growing cells) is most active and the bark is most easily peeled off of the trunk.

The inner bark (the phloem) conducts food from the crown to the other parts of the tree. As the tree produces additional phloem from the cambium, the old phloem cells become part of the outside bark. Cutting a strip of bark completely around the trunk is referred to as "girdling the tree." A girdled tree creates a continuous gap in the phloem and the tree will die of starvation.

The cambium, the actively growing part of the trunk, is between the phloem and the sapwood (xylem) on the inside. The cambium is most active in the spring, producing relatively large vessel elements of sapwood. The vessel elements are composed of conducting tissue and act like tiny pipelines, transferring food, minerals and water from the roots to the rest of the tree.

The vesssel elements are largest in the spring and become progressively smaller as growth slows throughout the summer and autumn and comes to a complete halt in the winter. The large vessel elements appear light coloured and become darker as the vessels get smaller. The alternating light and dark bands are known as "growth rings" and represent one growing season. An accurate age of the tree can be determined by counting all the growth rings present.

The growth rings can also help determine the growing conditions. Closely spaced growth rings indicate periods of low growth typically caused by competition for light and water from nearby trees. Wide growth rings indicate an abundance of light and moisture.

The transition of small growth rings to large growth rings often indicates the removal of competition in a process called "release." Release may happen several times in the life of the tree. Saplings undergo release when they grow above the lower canopy of grass and shrubs. Juvenile trees experience release by natural or mechanical thinning and selective cutting. Each time release occurs, it is accompanied by rapid growth indicated by widely spaced growth rings.

The sapwood carries water and minerals from the roots to the crown. The sapwood is most active just before bud break in the spring, and is usually the best time to extract the sap for secondary processing into syrup or sugar. Although we most commonly associate the sugar maple with sap production, other trees such as the Manitoba maple and the white birch have successfully produced syrup quality sap.

Mechanical abrasion by trail maintenance equipment.

The heartwood is sapwood that has lost the ability to conduct fluids and dissolved nutrients. As sapwood's vascular system begins to shut down with age, it begins to accumulate resins, gums, tannins and pigments, which cause the wood to darken. The heartwood is not much use to the tree except for giving the trunk additional strength, hence a tree may function perfectly well after the heartwood has rotted and left a hollow centre.

The sapwood and heartwood are usually the most commercially important part of the tree because the wood strength is sufficient for lumber and paper production. The cells in the heartwood and sapwood get their strength because the walls are composed primarily of cellulose fibres, which have enormous tensile and compressional strength.

This forester is touching a trunk scar resulting from frost damage.

The moisture content of the heartwood is usually considerably lower than that of the sapwood, the notable exceptions being the poplars. This moisture difference results in log cracking (checking) because of differential drying stresses. Frozen logs have little opportunity to dry out and therefore are not as susceptible to cracking. The drying and cracking of logs cut during the summer can be reduced by painting the butt ends.

Tree Trivia

The trunk of a tree only grows outward, never upward. A nail pounded into the trunk of a living tree will remain at the same height above the ground for the life of the tree.

The Crown

The crown is the upper part of a tree. It contains the branches and leaves. Its primary function is the production of carbohydrates and other related compounds used for plant growth. The crown also contains the apical bud, which is responsible for the upward growth of the tree.

Green leaves are able to use sunlight to chemically convert the carbon dioxide in the air into sugars and related substances in a process called "photosynthesis." These carbohydrates are then used by the tree for growth and reproduction.

The green leaf gets it colour from the chlorophyll present in the cell tissue. The chlorophyll heavily absorbs the wavelengths of light that are most effective in photosynthesis. Green light is not absorbed by the chlorophyll and hence is the reflected light we see.

The simplified chemical reaction of photosynthesis is shown below.

$$CO_2 + H_2O \xrightarrow[\text{Green Plants}]{\text{Light}} (CH_2O) + O_2$$

In other words,

$$\text{Carbon Dioxide} + \text{Water} \xrightarrow[\text{Green Plants}]{\text{Light}} \text{Carbohydrate} + \text{Oxygen}$$

The rate at which photosynthesis occurs is determined by the relative supplies of light, carbon dioxide (CO_2) and water. A shortage of one of these three relative to the others is called the "limiting factor."

For pioneer (shade intolerant) trees such as the poplars, pines and birches, often the limiting factor is water supply and the trees will show depressed growth (thin growth rings) during years of drought. In trees subsequent in the forest succession, often the limiting factor will be light until release occurs. CO_2 becomes the limiting factor only when adequate amounts of light and moisture are available.

CO_2 enters the green cells of the leaf through sausage-shaped openings called "stomata," and then through a series of interconnected air canals. The carbon dioxide then diffuses from the air canals into the cells where it is used in the photosynthetic process for the production of carbohydrates.

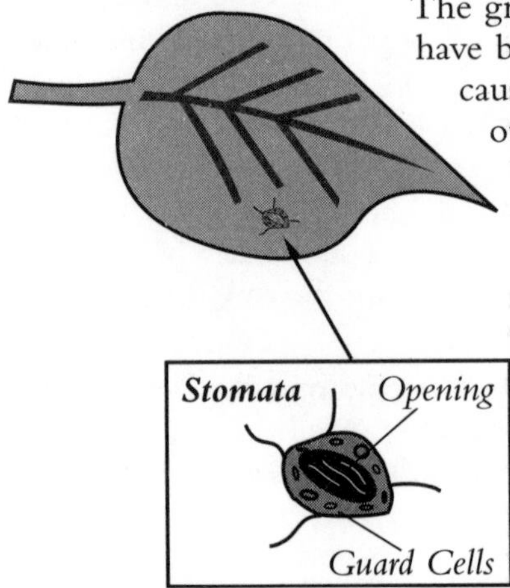

The underside of a leaf showing the stomata.

The greenhouse effect and global warming have been blamed on the CO_2 production caused by the burning of coal, oil and other fossil fuels. Because trees use CO_2 for food production, they have long been claimed to reduce the amount of greenhouse gas and, hence, reverse the global warming process.

A byproduct of photosynthesis is oxygen. The amount of oxygen a tree produces in a year depends on size, species and, of course, on the amount of macro and micronutrients present. The oxygen-producing capabilities of trees has led many people to call forest "the Earth's lungs."

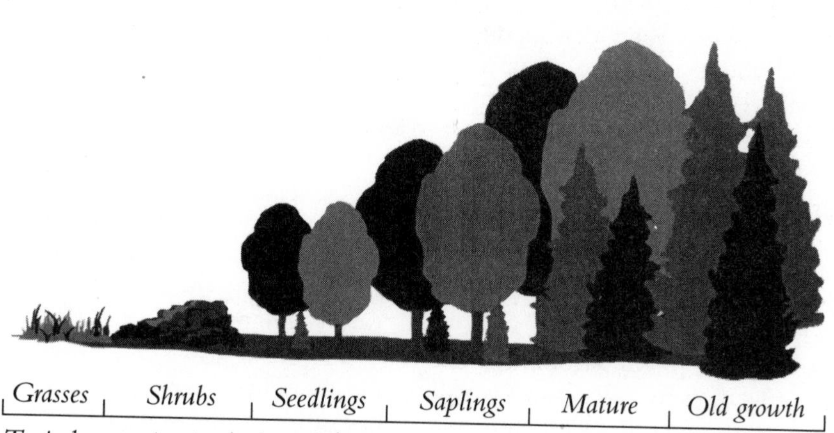

Typical succession in the boreal forest.

Tree Trivia

CO_2 can be the nutrient-limiting growth in well cared for house plants. CO_2 exists as a trace gas, comprising only about .03% (three parts per 10,000) of the atmosphere. A person's exhaled breath contains up to 5% of CO_2. House plants grow better when spoken to because they are using the additional CO_2 exhaled in the gardener's breath.

Forest Succession

Although a given woodland area appears to change little from year to year, forests are never static. They change continually in form and in composition in response to seasonal, climatic and nutritional factors. The day-to-day changes are barely noticeable except in the spring and autumn. The changes that occur in the forest over a long period can be classified into a number of stages based on species present, age, etc. This long-term change is called "forest succession."

A community of trees that exhibit common characteristics such as species, age, height or diameter is called a "stand," which is the basic management unit. A stand is as big or as small as the characteristics that define it.

Before the arrival of Europeans, most forest succession in the boreal forest was created by periodic natural disasters such as forest fires, wind blowdowns, and insect and disease outbreaks. These catastrophic disturbances removed the bulk of the canopy and the forest would start all over again with certain plants growing first and later being taken over by other species. Stands that escape these periodic natural disturbances for a long period of time develop into "old-growth forests."

The natural succession of a forest is usually divided into the following four stages, each identified by characteristic species: 1) pioneer (herb, shrub and seedling), 2) young forest (dominated by pine, poplar and birch), 3) mature forest (mixed wood, as above including spruce and balsam fir), and 4) climax old-growth forests (dominated by large diameter spruce and balsam fir).

The species of trees found in the various stages of forest succession are largely determined by the availability of light. Plants requiring full sunlight, such as poplar and pine, flourish immediately after the disturbance. As the shade intolerant tree species mature, reach old growth and die, they are replaced by shade tolerant species such as white spruce and balsam fir. If a catastrophic event does not occur for a long period of time, then the succession is culminated in a climax forest characterized by the growth of seedling and saplings that are the same species as the canopy.

The unmanaged boreal climax forest occurs after a relatively short period of time (250 years) compared to climax forests in coastal areas (over 500 years). Coastal forests tend to be longer lived because of climatic factors: the mild climate and increased rainfall promote the growth of longer lived tree species and reduce the incidence of forest fire.

Boreal and other forests in the Canadian interior often benefit when harvested by clearcut (SEE CHAPTER 5). The clearcut artificially simulates a naturally occurring catastrophic disaster such as fire. The forest regrowth will often occur in the same successional sequence as after a natural disturbance.

Knowledge about the basic components of trees and forests can be used in the application of silvicultural practices in a managed woodlot. Silviculture is Latin for "forest cultivation" and will increase forest productivity much the same way a tune-up will increase the fuel efficiency of a car or truck. Silvicultural practices can be used to alter the species present and the duration of any or all of the succession stages, resulting in a healthier, more productive woodlot.

Tree Trivia

Until 1985 in the province of Alberta, more forest was removed annually by the clearing of forests for farmland than all the logging operations combined!!!

Chapter Three

EVALUATING AN EXISTING WOODLOT

Economic Potential

To many landowners, a forested area is often valued for solely its recreational potential. In the prairie provinces it is frequently viewed as an obstacle to agricultural production. The idea that forested land is either of limited value or an expense has resulted in gross undervaluations of many farms and other land holdings.

There are essentially two considerations in determining the value of the trees on a given property: the total volume of wood and the value per unit volume of the wood.

The value per unit volume of wood on your property is determined by species, size, age, height, and local or personal demand. Forest value also increases as the woodlot manager performs more of the harvest operations.

The value of the trees depends on the end use. Timber used for pulp production has considerably lower value than wood sawn into boards or used in specialty products. Woodlot managers may increase the value of their timber by processing it into the end use that is most valuable in their area. Note that although the end use of the tree will determine the value of the tree standing, the cost of cutting, stacking and transport is similar regardless of tree species.

Value-added processes are discussed in more detail in Chapter 7.

Units of Measurement

One of the most important aspects of determining timber volume is understanding the units of measurement used in the forest industry. Timber volumes are measured in cords, cubic metres, board feet, metric tonnes and a whole host of in-between units. Below are some definitions and rules of thumb for converting the various units. I have used the most common industry measurements, which are not always metric and not always Imperial.

Roundwood refers to the solid volume of wood that occurs in a log or group of logs. It does not include the bark or the airspace that occurs in stacked wood. Roundwood is measured in cubic metres, that is, a cube 1 m high, 1 m wide, and 1 m long.

Because logs are round and taper towards one end, single tree volume tables have been developed to help determine roundwood volumes. The single tree volume tables are species- and area-specific and are available from local provincial forestry offices. See page 27 for sample calculations using these tables. Table 1 uses the single tree volume tables for aspen in sampling Region 8 for northeastern Alberta and do not necessarily apply elsewhere.

Roundwood can be scaled at the landing area using a scaling ruler, but most often volume is measured at the mill site on the basis of weight conversion factor. The conversion factor for deciduous trees is .95 tonne = 1 cubic metre. For coniferous trees the conversion factor is .87 tonne = 1 cubic metre.

The wood volume/weight relationships listed above are valid when the wood is cut and then weighed in a short period of time or in the winter when the wood remains frozen. If the wood is allowed to dry for a long period in the summer, these weight relationships are no longer valid, and because the trees lose moisture to evaporation, the volume/weight ratio will increase.

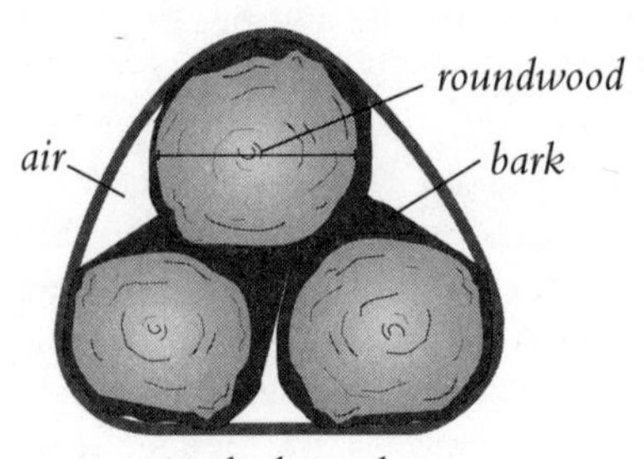

The difference between roundwood and stacked wood.

Stacked wood measurements are most often used in the firewood industry with the cord as a unit of measurement. A cord is defined as a pile of stacked wood containing 128 cubic feet of wood, bark and air. Typically this measurement refers to a stack of wood measuring 4 feet wide by 4 feet high by 8 feet long.

The standard board foot measurement.

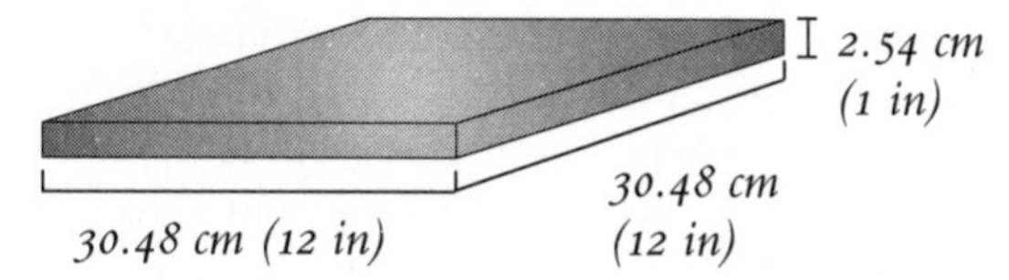

Because stacked wood includes bark and airspace between the pieces, the amount of roundwood in a given volume of stacked wood is considerably less. One cord of coniferous stacked wood contains only about 85 cubic feet of roundwood (solid wood). For deciduous wood the ratio is even lower, with only 71 cubic feet of roundwood in one cord of stacked wood.

Splitting the wood before stacking further reduces the amount of roundwood per cord by up to 20% depending on the species of tree.

The standard measurement for lumber is the board foot and has the abbreviation FBM (foot board measure). A board foot is the equivalent of a piece of wood 1 foot square and 1 inch thick. Lumber mills rate their production in units of 1,000 board feet (mFBM). For example, a mill may have a daily production of 60,000 board feet per day (60 mFBM/d) and a yearly production of 2 million board feet per year (2 mmFBM/y).

Some confusion arises when the differences between nominal and actual sizes are taken into consideration. The nominal lumber size is the dimension of a rough green board before planing and edging. The actual size of the lumber is the size of the board after it has been dried, planed and edged.

For example, most retail lumber yards sell boards using their nominal size. If you purchase a two by six 16 feet long (nominal size), you will find if you measure that board that the actual size is 5.5 inches by 1.5 inches. The actual size is always smaller than the nominal size because of edging and planing. The metric measurement is the actual size converted to metric units.

To clarify the situation, consider a two by six, 16 feet long. This board has nominal dimensions of 2 inches by 6 inches by 16 feet long (16 board feet) before processing. This is called the board's "nominal size." This same board after drying, planing and edging has a standardized actual size of 1.5 inches by 5.5 inches by 16 feet long (11 board feet) but it is still called and sold by retailers as a two by six, its nominal size. The metric size is the actual size (1.5 inches x 5.5 inches x 16 feet long) converted into metric units, in this case a 38 mm x 140 mm x 4.9 m.

In other words:

Example Board =2 inches x 6 inches x 16 feet

•Nominal = $\frac{2" \times 6" \times 16'}{12 \text{ square inches}}$ = 16 FBM (board feet)

•Actual = $\frac{1.5" \times 5.5" \times 16'}{12 \text{ square inches}}$ = 11 FBM

•Metric = 38 mm x 140 mm x 4.9 m = 0.026 cubic metres

Lumber volumes can be estimated from roundwood volumes. Obviously it is necessary to consider the waste wood such as the sawdust, shavings and chipwood, which are byproducts of converting a round log into square boards.

Sawmill Wood Waste Analysis 1989-90
Weighted average of all major sawmills

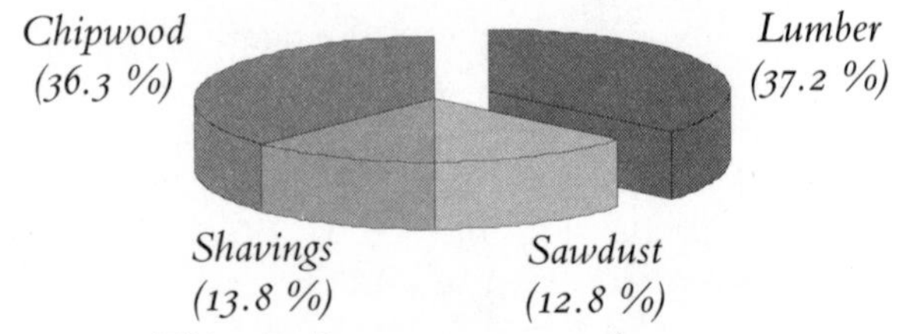

This pie diagram represents the average lumber recovery from roundwood in Alberta sawmills.

For conifers, the average percent recovery of lumber from roundwood is 37%, therefore 1 cubic metre of roundwood will produce an average of .37 cubic metres, or 233 FBM (nominal board feet) of lumber. No commercial lumber recovery figures for aspen are available. From personal experience, recovery is estimated at 50% of conifer recovery or about 116.5 FBM/m^3 roundwood.

Estimating Timber Volume

Once you have a reasonable understanding of the units of measurement, there are three ways of determining timber volume and value on a given parcel of land. The first is letting the forest company interested in buying your timber do the work, the second is to hire a forestry consultant, and the third is to do the timber volume calculations yourself.

Allowing a sawmill or pulpmill company representative to evaluate your timber can be the best option when it is impractical to do it yourself. The timber company will often develop a logging plan, arrange for logging contractors and assist in reforestation. It may also be able to get a cash advance on timber to be logged in the future. This situation works best when several companies are competing for your timber.

Timber cruising is fun, and the whole family can be involved.

There may be disadvantages to allowing the timber company to evaluate your woodlot. Large timber companies make their profit by moving large volumes of fibre. They usually will not look at volumes less than one logging truck load (45 cubic metres) which amounts to about 200 trees. Since it is the nature of business to make a profit, obviously no matter how honest the company representative is, he or she will be looking to make the best deal for the company. Also, the amount of time spent with you developing a forest management plan is an expense to the company and often this expense is passed on to the landowner in the form of a lower timber price.

Hiring a consultant is usually only economically viable for landowners who have large forested areas or highly valued species of trees. The consultant will provide an unbiased assessment of your wood volume and the value of the trees. His or her knowledge of mill margins, and local and international price projections may put you in better bargaining position with the timber buyer and give you a better idea how much wood to sell when.

The disadvantage of hiring a consultant is primarily the capital cost.

Doing volume estimates yourself has a number of advantages. First it is usually much cheaper, especially for small stands of timber, and it allows you to familiarize yourself with your woodlot. By going out and making the assessment yourself, you have an opportunity to observe the general health of the stand, see improvements to be made and select the best locations and logging methods that suit your style of management.

There are also many intangible benefits to doing volume estimates yourself. Number one, it's fun. Measuring trees does not have to be done alone. In fact, the job can be done more enjoyably and quickly if the whole family joins in. Second, forest measurement is a good way of getting a little fresh air and some exercise. It is also a good excuse to get away: instead of telling people you're going to kill some time with a walk with the dogs in the woods, you can tell them you're going out "timber cruising"!

On very small woodlots, such as those in urban areas, wood volume can be obtained simply by going out and counting and measuring all of the trees in the woodlot. For woodlots comprising a hectare (2.2 acres) or more, measuring all of the trees is often just not practical. Accurate volume estimates can be obtained by first determining the treed area that exists and then by sampling small areas of the woodlot and extrapolating the results for the entire area.

Estimating treed areas requires a recent air photo of the parcel of land you are interested in. Air photos have been taken of most of the land areas in Canada and can be obtained from the provincial or territorial government maps branch. Unless a very large parcel of land is being evaluated, try to get the biggest enlargement as possible. Usually a 1:5000 scale is sufficient.

A stereoscope makes it possible to get overlapping stereoscopic air photos. This type of air photo shows apparent relief (three dimensions) when viewed with stereoscopic glasses and is used to determine tree height, valley contours, etc., when timber cruising is not feasible. For the purpose of the woodlot owner, however, stereoscopic air photos are not usually necessary.

An air photo's accuracy diminishes with time, so check the date they were taken. If the air photo is more than a few years old and the land is unfamiliar, it may be a wise to fly over or ground cruise the area of interest. Natural disasters such as spruce budworm infestation, fire or blowdown can dramatically change the economic potential of a forested area in a short time.

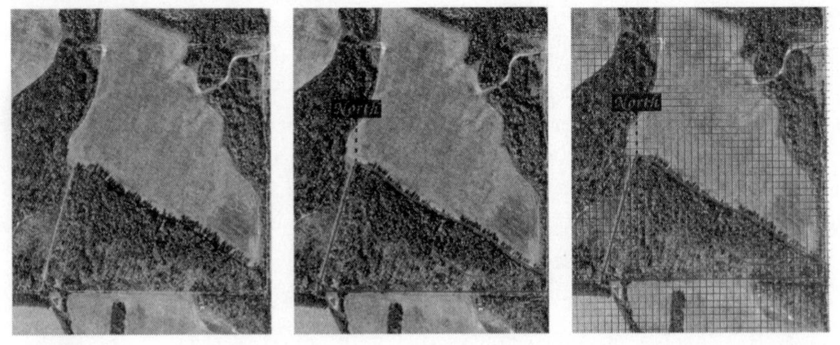

The above 1:5000 air photos were reduced to appear in this book. The air photo (left), taken in the summer, shows an almost continuous deciduous canopy. The air photo is then divided into homogeneous units (HUs, or "hueys") based on tree density (middle). In this example, aspen is the predominant species, so the hueys will be determined by tree density. A line has been drawn around the northern area, which appears to have a relatively high tree density and is denoted as HU #1. Calculate the area of HU #1 by tracing its outline on graph paper (right).

Another consideration when evaluating the economic potential of forest land from air photos is the time of year they were taken. Air photos taken in the summer may show a deciduous canopy but may hide a potentially valuable conifer understorey. Air photos taken in the early spring, late autumn and winter will reveal the conifers but it is impossible to distinguish between dead standing and living deciduous trees without their leaves.

First determine the orientation of the line and distance on the air photo.

By using the air photo, regions within the woodlot with similar tree density, age and species (conifer versus deciduous) can be identified and subdivided into Homogeneous Units (HUs, or "hueys" for short). Hueys that cover a large acreage can be further subdivided by geographical boundaries such as fields, creeks and ridges to add accuracy to the wood volume estimates.

Each huey should be given an identifier such as a number (HU #1, HU #2, etc.) or name (Enchanted Forest, Wendy's Woods, etc.) to separate the various units on the map and avoid confusion on the measurement sheets.

To calculate the area of the HU,

1) Trace the outline of the HU on a piece of graph paper
2) Find the area of each square
3) Count the total number of squares in the HU
4) To find the total area, multiply the area of the squares by the total number of squares contained in the HU. For example:

Scale of air photo = 1:5000, which means 1mm = 5 m
1 small square = 2.5 mm long by 2.5 mm wide
which represents an area = 12.5 m long by 12.5 m wide
therefore 1 small square = 156.25 m^2

HU#1 = Sum of small squares in HU x m^2 per small square
= 160 small squares x 156.25 m^2 per small square
= 24,960 m^2 total area rounded to the nearest decimal
= 2.5 hectares x 2.47 acres per hectare rounded off
= 6.2 acres

Line plot width is determined by holding out a measured stick at arm's length.

Forest Sampling

Once the size of the HU has been calculated, it is then necessary to estimate the volume of wood per unit area by taking tree height and diameter measurements in small sample areas within the HU. With these tree measurements, tree volume can be calculated in the sample area and then extrapolated to the whole HU. At least one but preferably more sample plots should be measured per HU. Theoretically, more sample plots result in more accurate volume estimates, but going overboard on the number of sample plots defeats the purpose of sampling. There are many methods of sampling stands of timber, but the most commonly used is the line plot.

The line plot consists of traversing a straight line from one edge of a sampling area to the other and measuring all trees that occur within 3 m (10 feet) of the line. First, determine the line direction and distance on the air photo. Determine line orientation by aligning north on the air photo with north on the compass and estimating angle of deviation. Note any topographic features, large trees or man-made structures that could be used as reference points in the field. Distance can be measured by using a ruler and multiplying by the scale of the map.

Once out in the sampling area, identify the reference points and sample in a straight line between them. A compass can be used in featureless areas where no reference points are available. The width of the sample plot can be determined by holding a measured stick arm's length and recording every tree that comes in contact with the stick or body.

Measure the height and diameter of the trees in each sampling area. Because tree diameter is by far the easiest measurement to make, it is usually only necessary to measure the height of a few trees because trees with similar diameter can be assumed to have similar heights providing they are the same species.

Height can be measured in several ways. The easiest way is to find a fallen tree or falling a representative tree and measuring the length by using a tape measure or counting paces. Measure the diameter of the fallen tree as well so the two can be correlated with other trees.

Calculating tree height using triangulation.

Another way to determine tree height is by triangulation. A right angle triangle has the vertical component equal to the horizontal component, that is, side A = side B. If you create a visual right angle triangle, you can estimate the height of the tree by measuring the distance from the tree.

To create a visual triangle, take a yardstick and hold it upright in your hand. The amount of the yardstick above your hand should equal the length of your reach so side A = side B. With the arm stretched out, adjust the height of your fist so that it lines up with the base of the tree. Then move away from the tree until the line of sight with the top of the yardstick intersects with the top of the tree, and stop at this position. To measure the height of the tree, count the number of paces back to the tree.

A easy rule to remember when using your pace to measure distances is that your pace (two normal steps) is the same length as your height. So if you are 2 m tall (6.6 feet), your average pace will be 2 m.

In an even aged stand, most of the canopy trees will be a similar height. Only a few tree height measurements are necessary for the whole stand. In uneven stands more tree height measurements will usually be necessary.

Diameter at breast height (dbh) measurements are also very simple to make. Because breast height varies with the person doing measuring, dbh has been standardized to the tree diameter that occurs about 1.5 m (4.5 feet) off the ground.

A diameter tape has the diameter calculated directly from the circumference.

It usually necessary to measure only the commercial trees in a stand. Non-commercial trees such as dead standing trees and trees with a dbh of less than 10 cm (4 inches) are not usually measured. The easiest way to make diameter measurements is to buy a forestry diameter tape. Although the tape is placed around the circumference of the tree trunk the tape is calibrated so that units of diameter can be read directly. If a diameter tape is not available, measure the circumference with a cloth (sewing) measuring tape and divide by a factor of 3.14 to get the diameter value. If the measurements are taken when you are alone, use a pushpin to hold the tape in place on trees with a large diameter. Because the formula relating circumference (C) to diameter (D) is:

C = (pi) x D

where (pi) = 3.14 *then* C = 3.14 x D
therefore the circumference divided by 3.14 *gives the diameter, or* C/3.14 = D.

To calculate the volume of timber in HU #1:

1) Find the direction of the traverse from the air photo
2) Calculate distance of traverse on air photo
3) Find traverse starting point and direction in field
4) Count paces or use hip chain to determine distance in field
5) Use measured stick to determine width of traverse
6) Measure height and diameter of all trees that can be touched by the measured stick along the traverse
7) Record the height and diameter at breast height measurement of all commercial trees. Calculate total roundwood volume by multiplying the number of trees in a diameter height class by the single tree volumes in Table 1. (SEE SAMPLE, PAGE 27)
8) Calculate roundwood density of sample traverse (SEE SAMPLE, PAGE 28)
9) Calculate total volume of HU. (SEE SAMPLE, PAGE 28)
10) Calculate value of timber in HU #1 (SEE SAMPLE, PAGE 28)

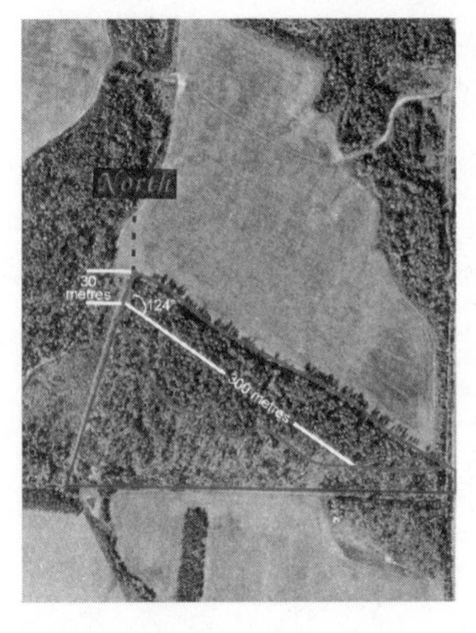

Orientation and distance of traverse.

Sample Record of dbh, number of trees and total roundwood volume:

Date: March 23, 1995
HU: HU #1
Species: Aspen
**These figures are derived from* Table 1, *below.*

DBH (cm)	*Number of trees*	*Volume / Tree** (m^3)	*Total Volume* (m^3)
Tree Height: 12 m			
10			
12	2	0.05	0.10
14			
Tree Height: 14 m			
16	5	0.11	0.55
18	11	0.14	1.54
20	7	0.17	1.19
Tree Height: 16 m			
22	9	0.23	2.07
24	7	0.28	1.96
26	3	0.32	0.96
28	2	0.37	0.74
30	3	0.42	1.26
Total	**49 trees,** 10.37 m^3 roundwood in sample traverse		

ALBERTA FOREST SERVICE - PHASE 3 INVENTORY
SINGLE TREE VOLUME TABLE 1984
VSR 8 GROSS MERCHANTABLE CUBIC METRE VOLUME ASPEN
7.0 CM TOP DIB 0.30 M STUMP HT

DBH CM	HEIGHT IN METRES 4	6	8	10	12	14	16	18	20	22	24	26	28	30	32	34	36
2	0.00	0.00	0.00	0.00	0.00	0.00	0.00	0.00	0.00	0.00	0.00	0.00	0.00	0.00	0.00	0.00	0.00
4	0.00	0.00	0.00	0.00	0.00	0.00	0.00	0.00	0.00	0.00	0.00	0.00	0.00	0.00	0.00	0.00	0.00
6	0.00	0.00	0.00	0.00	0.00	0.00	0.00	0.00	0.00	0.00	0.00	0.00	0.00	0.00	0.00	0.00	0.00
8	0.00	0.00	0.01	0.01	0.01	0.01	0.02	0.02	0.02	0.02	0.03	0.03	0.03	0.04	0.04	0.04	0.04
10	0.01	0.01	0.02	0.02	0.03	0.03	0.04	0.05	0.05	0.06	0.07	0.07	0.08	0.09	0.10	0.10	0.11
12	0.01	0.02	0.03	0.04	0.05	0.06	0.07	0.08	0.09	0.10	0.11	0.12	0.13	0.15	0.16	0.17	0.18
14	0.02	0.03	0.04	0.05	0.07	0.08	0.10	0.11	0.13	0.14	0.16	0.17	0.19	0.21	0.22	0.24	0.26
16	0.02	0.04	0.05	0.07	0.09	0.11	0.13	0.15	0.17	0.19	0.21	0.23	0.25	0.27	0.29	0.32	0.34
18	0.03	0.05	0.07	0.09	0.11	0.14	0.16	0.18	0.21	0.24	0.26	0.29	0.32	0.34	0.37	0.40	0.43
20	0.03	0.06	0.08	0.11	0.14	0.17	0.20	0.23	0.26	0.29	0.32	0.35	0.39	0.42	0.45	0.49	0.52
22	0.04	0.07	0.10	0.13	0.16	0.20	0.23	0.27	0.31	0.35	0.38	0.42	0.46	0.50	0.54	0.59	0.63
24	0.05	0.08	0.12	0.15	0.19	0.23	0.28	0.32	0.36	0.41	0.45	0.50	0.54	0.59	0.64	0.69	0.74
26	0.05	0.09	0.13	0.18	0.22	0.27	0.32	0.37	0.42	0.47	0.52	0.58	0.63	0.69	0.74	0.80	0.85
28	0.06	0.11	0.15	0.20	0.26	0.31	0.37	0.42	0.48	0.54	0.60	0.66	0.72	0.79	0.85	0.91	0.98
30	0.07	0.12	0.17	0.23	0.29	0.35	0.42	0.48	0.55	0.61	0.68	0.75	0.82	0.89	0.97	1.04	1.11
32	0.08	0.14	0.20	0.26	0.33	0.40	0.47	0.54	0.61	0.69	0.77	0.85	0.92	1.01	1.09	1.17	1.25
34	0.09	0.15	0.22	0.29	0.37	0.44	0.52	0.60	0.69	0.77	0.86	0.94	1.03	1.12	1.21	1.31	1.40
36	0.10	0.17	0.24	0.32	0.41	0.49	0.58	0.67	0.76	0.86	0.95	1.05	1.15	1.25	1.35	1.45	1.55
38	0.11	0.19	0.27	0.36	0.45	0.54	0.64	0.74	0.84	0.95	1.05	1.16	1.27	1.38	1.49	1.60	1.71
40	0.12	0.20	0.30	0.39	0.49	0.60	0.70	0.81	0.92	1.04	1.15	1.27	1.39	1.51	1.63	1.76	1.88
42	0.13	0.22	0.32	0.43	0.54	0.65	0.77	0.89	1.01	1.13	1.26	1.39	1.52	1.65	1.79	1.92	2.06
44	0.14	0.24	0.35	0.47	0.59	0.71	0.84	0.97	1.10	1.24	1.37	1.51	1.65	1.80	1.94	2.09	2.24
46	0.15	0.26	0.38	0.51	0.64	0.77	0.91	1.05	1.19	1.34	1.49	1.64	1.79	1.95	2.11	2.27	2.43
48	0.17	0.28	0.41	0.55	0.69	0.83	0.98	1.13	1.29	1.45	1.61	1.77	1.94	2.11	2.28	2.45	2.63
50	0.18	0.31	0.44	0.59	0.74	0.90	1.06	1.22	1.39	1.56	1.73	1.91	2.09	2.27	2.45	2.64	2.83
52	0.19	0.33	0.48	0.63	0.80	0.96	1.14	1.31	1.49	1.68	1.86	2.05	2.24	2.44	2.64	2.84	3.04
54	0.21	0.35	0.51	0.68	0.85	1.03	1.22	1.40	1.60	1.79	1.99	2.20	2.40	2.61	2.82	3.04	3.25
56	0.22	0.38	0.55	0.72	0.91	1.10	1.30	1.50	1.71	1.92	2.13	2.35	2.57	2.79	3.02	3.25	3.48
58	0.23	0.40	0.58	0.77	0.97	1.17	1.38	1.60	1.82	2.04	2.27	2.50	2.74	2.98	3.22	3.46	3.71
60	0.25	0.43	0.62	0.82	1.03	1.25	1.47	1.70	1.94	2.17	2.42	2.66	2.91	3.16	3.42	3.68	3.94
62	0.26	0.45	0.66	0.87	1.10	1.33	1.56	1.81	2.05	2.31	2.56	2.83	3.09	3.36	3.63	3.91	4.18
64	0.28	0.48	0.70	0.92	1.16	1.41	1.66	1.91	2.18	2.44	2.72	2.99	3.27	3.56	3.85	4.14	4.43
66	0.30	0.51	0.74	0.98	1.23	1.49	1.75	2.02	2.30	2.59	2.87	3.17	3.46	3.76	4.07	4.38	4.69
68	0.31	0.54	0.78	1.03	1.30	1.57	1.85	2.14	2.43	2.73	3.03	3.34	3.66	3.97	4.30	4.62	4.95
70	0.33	0.57	0.82	1.09	1.37	1.65	1.95	2.25	2.56	2.88	3.20	3.52	3.86	4.19	4.53	4.87	5.22
72	0.35	0.60	0.86	1.15	1.44	1.74	2.05	2.37	2.70	3.03	3.37	3.71	4.06	4.41	4.77	5.13	5.49
74	0.36	0.63	0.91	1.20	1.51	1.83	2.16	2.49	2.84	3.18	3.54	3.90	4.27	4.64	5.01	5.39	5.77
76	0.38	0.66	0.95	1.26	1.59	1.92	2.27	2.62	2.98	3.34	3.72	4.09	4.48	4.87	5.26	5.66	6.06
78	0.40	0.69	1.00	1.33	1.66	2.01	2.38	2.74	3.12	3.50	3.90	4.29	4.69	5.10	5.52	5.93	6.36
80	0.42	0.72	1.05	1.39	1.74	2.11	2.49	2.87	3.27	3.67	4.08	4.49	4.92	5.34	5.78	6.21	6.66

Table 1. Aspen single tree volumes for Region 8 in northeastern Alberta. (Courtesy Alberta Environment Protection.)

Sample calculation of roundwood density in sample traverse

Sample density = roundwood volume/area of traverse

Distance of traverse = 300 m

Width of traverse = 4 m

Area of traverse = 300 m x 4 m = 1200 m^2

Total volume of trees in sample area = 10.37 m^3

Density of roundwood in sample area = $\frac{10.37\ m^3}{1200\ m^3}$ =.000865 m^3 roundwood

Sample calculation of total volume of roundwood in HU #1

Total volume = sample density x area of HU #1

= .000865 m^3 roundwood/m^2 x 25000 m^2 (2.5 hectares)

= 216.25 m^3 roundwood

= 216 m^3 roundwood (rounded off)

Sample calculation of value of timber in HU#1

As standing timber:

216 m^3 roundwood x .85 tonnes/m^3 = 183.6 tonnes

183.6 tonnes x $2.00/tonne = $367.20

As decked timber (piled delimbed logs):

183.6 tonnes x $11.00/ tonne = $2019.60

As firewood:

216 m^3 roundwood x .5 cords firewood/m^3 roundwood = 108 cords

108 cords x $50.00/cord = $5400.00

As lumber:

216 m^3 roundwood x 116.5 board feet (FBM)/m^3 roundwood

= 25164 FBM [25,000 board feet (mFBM)]

25 mFBM x 300.00/mFBM = $7500.00 + residual material

Agroforestry

Agroforestry is the combined practices of agriculture and silviculture for mutual benefit. Agroforestry can take many forms including 1) orchards, 2) tree farms, 3) shelterbelts, 4) silvipasture, or 5) some combination of the four.

Orchards

Many types of orchards flourish in the fruit-growing regions of B.C., Ontario, Quebec and the Maritimes where a mild microclimate is suitable for growing everything from apples to peaches. Most farms that lie outside the fruit belts can still grow orchards of winter-hardy fruit. Private and government research centres have both developed many varieties of apples, apricots, plums, cherries and pears to withstand even the harshest climates.

Other orchards cultivate fruit that naturally occurs in the wild. Demand for fruit such as saskatoons, pincherries and chokecherries is on the rise as the public becomes exposed to and develops a taste for these types of fruit. Saskatoons can be enjoyed right off the tree, but most other types of berries require secondary processing in jams, jellies and syrups.

Tree Farming

Tree farming is becoming popular in many parts of the country. Tree farming is similar to traditional crop production, but it is not as simple as planting a inexpensive seedling and selling a high priced tree in a few years. Most tree farms are extremely labour and capital intensive. Canadian tree farms usually grow either nursery trees for ornamental purposes or Christmas trees.

Nursery tree production can be fun and profitable especially as a small-scale hobby. Usually the nursery is associated with a greenhouse that can grow trees to the seedling stage. The seedlings are then planted outside where they are fertilized and irrigated until they become a suitable size for sale, usually within a few years. Tree farming has tremendous potential as many provincial governments are pulling out of the tree growing business.

Marketable Christmas trees.

Christmas tree growing is also another common type of agroforestry. Although the market is saturated in many parts of the country, there are still opportunities for local growers who can produce high quality Christmas trees. Scots pine, Douglas fir and balsam fir are the most popular species to grow, although just about any conifer will do.

After stand establishment, the most important aspect of Christmas tree growing is to create a marketable inverted cone-shaped tree. Periodic shearing and pruning beginning at the third or fourth year of tree growth and continuing until harvest is necessary to properly shape the tree. Initial pruning operations attempt to correct imperfections such as broken or double leaders. Subsequent operations are directed at filling in "goosenecks" or gaps in foliage and cutting a "handle" at the base of the tree for easy handling and insertion into the stand.

Christmas trees can be marketed either as a U-cut operation, where the potential buyer walks through the stand and selects a tree, or as a wholesaler to a Christmas tree lot. When selling wholesale, wrap the compressed tree branches with twine using a tree baler to increase the carrying capacity of the transportation vehicle by two and a half times.

Shelterbelts

Shelterbelts are beneficial in just about every aspect of farming. Shelterbelts have been proven to reduce heat loss in homes, increase returns in livestock and crop operations, reduce soil erosion and increase wildlife populations. The list is virtually endless.

Many shelterbelts are planted on field boundaries to show differences in ownership. Properly planted shelterbelts can dramatically increase crop yields, especially in areas where lack of moisture is the limiting factor in crop production. The shelterbelt's main function is to reduce wind speed, which in turn traps drifting snow, prevents snow from being carried from the place it has fallen in the winter and slows plant and soil desiccation in the summer.

Topsoil erosion like this could have been prevented by proper planting of shelterbelts.

Often a single shelterbelt along field margins is not sufficient to benefit an entire field and it is necessary to plant additional intermediate parallel shelterbelts. A rule of thumb to determine lateral

Snowmelt from the snowdrifts trapped behind the shelterbelt will provide water for spring crop growth.

shelterbelt effectiveness is to multiply the height of the shelterbelt by ten. For example, a 10 m (33 foot) high shelterbelt will have a wind shadow of 30 m (100 feet).

Shelterbelt effectiveness in trapping snow also depends on several other factors including orientation and presence of stubble and other shelterbelts, roads or other obstacles nearby. For example, shelterbelts placed 100 m (330 feet) apart will trap less snow per shelterbelt (although the total snow trapped will be greater) than a single isolated shelterbelt.

Trapped snow provides additional soil moisture in the spring for planting. The additional snow cover also acts as an insulating layer to autumn-seeded field and horticultural crops. Winter wheat, fall rye and specialty crops such as strawberries and tree seedings will not be as susceptible to stress or winterkill from extreme winter temperatures and will emerge from dormancy in a healthier condition.

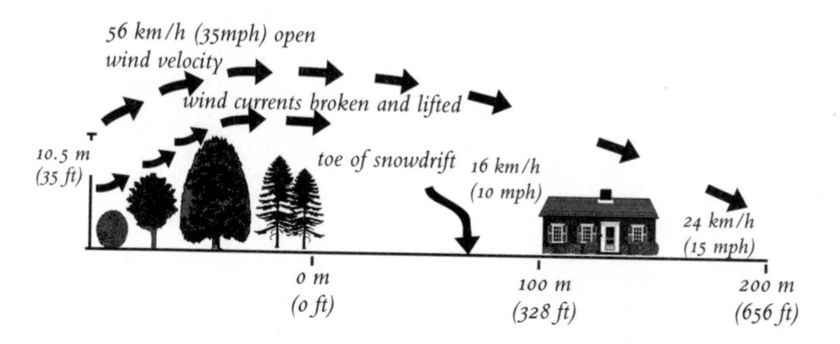

A dense windbreak, properly designed and correctly located, will effectively reduce wind velocity and control snowdrifting. The rural dwelling sheltered from the wind will be more comfortable and will have up to 35% lower heating costs. (Adapted from PFRA.)

Recommended species and spacing of trees within rows for shelterbelt configuration. (Courtesy PFRA.)

1.0 m (3 ft)	*2.4 m (8 ft)*	*2.4 m (8 ft)*	*3.0 m (10 ft)*
Vilosa Lilac *Buffaloberry* *Chokecherry*	*Willow* *Poplar*	*Green Ash* *Manitoba Maple* *Willow*	*Colorado Spruce* *White Spruce* *Scots Pine*
0.3m (1 ft) *Caragana*			

Growing crops also benefit from shelterbelts through increased soil moisture levels, daytime temperatures, humidity and night-time carbon dioxide levels. Worldwide studies indicate that shelterbelts can increase crop quality and maturity can be hastened by up to four or five days.

Shelterbelts can be an advantage to harvested crops as well. Light crops such as canola, flax and mustard can be blown from the swath on blustery autumn days, resulting in seed loss and additional difficulty in harvesting.

Shelterbelts also reduce heat loss from buildings and livestock. Air movement increases convective heat loss and is the main reason we feel colder on windy days. The wind chill factor (SEE TABLE 2) shows the relationship between wind speed and heat loss. By reducing the wind speed with a shelterbelt, home heating costs can be reduced and livestock performance will be increased.

Windspeed (Km/hour)	*Actual Temperature Reading °C*					
	10	0	-10	-20	-30	-40
	Apparent Temperature Reading °C					
calm	10	0	-10	-20	-30	-40
10	9	-1	-12	-23	-34	-45
20	4	-8	-20	-30	-42	-55
30	2	-12	-25	-37	-49	-65
40	0	-13	-27	-42	-55	-72
50	0	-14	-30	-44	-58	-77
60	-1	-15	-31	-46	-63	-79

Table 2: Wind Chill Factor

Despite the benefits shelterbelts provide, remember that trees take land out of crop production and compete with crops for light, moisture and nutrients. Another disadvantage is the short-term expense of money and time for shelterbelt establishment, and the relatively long time in reaping the maximum rewards.

The economic returns of a properly designed shelterbelt far outweigh any negative impacts. Shelterbelts have been shown to increase crop yields by up to 15% while only taking up 5% of the land area resulting in a net increase in yield of 10% not including the value of trees if harvesting is so desired. Shelterbelts also have many intangible benefits such as wildlife habitat and farm beautification. A suggested shelterbelt configuration is shown opposite. For more information on shelterbelts and their establishment, contact the local Prairie Farm Rehabilitation Administration (PFRA) office.

Silvipasture

Silvipasture is the practice of grazing animals in a forested area. Many private forest product companies recognize the advantages of integrating livestock into their reforestation programs. Flocks of sheep are most commonly used for vegetation control in areas newly planted to conifers (SEE CHAPTER 8), but other small-framed animals such as dexter cattle and goats may also be used.

Livestock can benefit the woodlot by grazing grasses and low shrubs. By keeping the grass and other brushy vegetation cropped low to the ground, the danger of fire can be greatly reduced and newly planted seedlings can attain maximum growth with minimum competition. Nutrients are returned to the forest in the form of manure and urine.

Grazing livestock is not suitable for all woodlot applications, however. Livestock should be excluded from newly planted deciduous stands, as the livestock will browse on the new leafy growth. Shallow-rooted trees such as poplar and spruce are easily damaged from the hoofs of larger animals such as cattle and horses. Trails and haul roads may be damaged by

This is a poor example of a silvipasture area. It is unsafe for cattle, and overcrowding has caused root damage to the trees.

The final resting place for many urban forests is usually the local landfill.

large numbers of individuals travelling one behind the other and creating "cow paths." Livestock manure may reduce the forest aesthetics or degrade the water quality of the forest pond or stream. When considering silvipasture as part of the woodlot operation, consult the provincial agriculture or forestry official in your area.

Urban Forestry

Although urban forestry sounds a bit like an oxymoron, great potential exists for both forest propagation and harvesting in our towns and cities. Many individuals and municipalities are reaping the financial and environmental rewards of operating a forestry business in the heart of suburbia.

Many people think of commercial "forests" as a few large areas with thousands of trees on each area. Urban woodlots, on the other hand, can be viewed as thousands of small areas with only a few trees on each.

Urban forests differ in other ways. They tend to be composed of valuable, well-maintained trees. Because most urban residences have streets and back alleys connecting them, large equipment used for tree removal and road construction is unnecessary.

Urban forests also tend to be valued much more highly for their visual, aesthetic and recreational aspects while their economic potential has been largely ignored. Often dead or dying trees are removed and hauled to the municipal landfill.

Growing an urban woodlot can serve multiple purposes above and beyond beautifying the property. Trees, because of their large capacity for nutrients and water uptake, can efficiently use a large portion of domestic waste water. The nutrient-rich water coming from the clothes washer, showers and baths, called "greywater," is essentially odourless and can be carried through separate plumbing to irrigate trees and other vegetation surrounding the house. Toilet and dishwasher water flows through regular plumbing into the public sewer system. In the winter, a valve system can redirect the greywater into the public sewer system.

Fields of trees grown using waste water irrigation overlook the city of Vernon, B.C.

The city of Vernon in the beautiful Okanagan valley in southern B.C. has taken urban forestry one step further. Vernon uses its entire waste water discharge of four million cubic metres (900 million gallons) to irrigate 300 hectares (740 acres) of agricultural land under intensive management, 400 hectares (988 acres) of wild rangelands and approximately 15 hectares (37 acres) of conifer and hybrid poplar. Before the irrigation program, the entire volume of waste water was dumped into Lake Okanagan.

The results of the irrigation program have been nothing short of spectacular: the water quality of Lake Okanagan has greatly improved around Vernon; the trees add beauty and help moderate the local climate; and marketable trees are produced in less than 15 years!

Growing trees also act as air conditioners and purifiers. The trees remove some of the carbon dioxide produced by motor vehicles to facilitate photosynthesis, and as a byproduct, release oxygen, which we breath, and water vapour, which cools the air near the tree. The release of oxygen and water vapour are two reasons it is so much more refreshing to sit in the shade of a tree than that of a building.

Separate plumbing systems can use domestic "greywater" to water trees and greatly reduce the burden on public water treatment facilities.

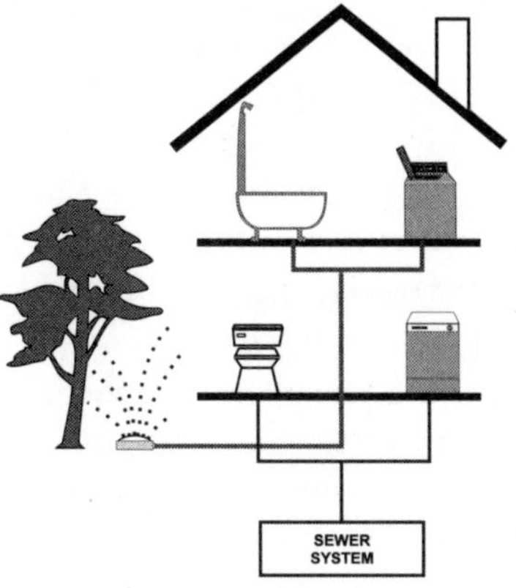

Dr. Mike Carlson beside a 12-year-old hybrid poplar tree near Vernon, B.C. that is almost ready for harvest.

Trees, as mentioned earlier in this chapter, serve to reduce wind speed resulting in lower heating and air conditioning costs and providing a sheltered growing area for the lawn and garden. Deciduous trees also can be placed in strategic positions to shade the house in the summer while allowing the sun's warming rays through in the winter (SEE OPPOSITE).

Harvesting urban forests is usually restricted to removing diseased, dying or dead trees or trees that interfere with overhead power lines or pose a threat to people or property. Occasionally larger areas may be harvested especially if a large number of trees have blown down after a windstorm or other natural disaster. Several individuals have profited from removing boulevard trees affected by dutch elm disease.

Harvesting is not recommended for the weekend lumberjack and should be left to individuals or companies with the proper training and equipment. A wood supply can be obtained from a tree removal company or purchased from the municipality or landowner after the tree is down. In many instances either the company or the landowner will give you the tree to save "transportation to the dump" costs.

Once a wood supply has been secured, numerous opportunities exist in urban areas for the manufacture and marketing of wood and wood products. Ornamental species commonly found in towns and cities may provide highly valued wood. Value can be increased when the wood is manufactured into tables, bowls, lamps and carvings. Even branches from a pruning operation can be made into decorative wreaths and furniture.

The portable bandsaw mills have opened a whole new dimension in urban wood manufacture. Bandsaw mills are light, quiet and portable. They can be stored in the garage and moved to where lumber is to be sawn, or they can be used in your backyard. These mills produce less sawdust and waste wood than traditional circular saws so even cleanup is reduced. Bandsaw mills are particularly adept at making valuable specialty cuts.

Deciduous trees provide shade in the summer and allow the sun to warm the house in the winter.

Urban forestry operates on many of the same principles and offers many of the same benefits as rural forestry does without the corresponding land base. Urban woodlot management can be fun and profitable for the whole family. If a change is as good as a holiday, many people chained to a desk can enjoy a holiday every day in their own urban forest.

Tree Trivia

Many cities located in earthquake zones encourage residents to build wood-framed houses. Wood, because of the cellulose fibres in the cell walls, has excellent tensile and shear strength, making wood-framed houses more likely to withstand the effects of an earthquake than houses built from brick or concrete.

These elm cants, harvested from boulevard trees, will soon be sawn into valuable lumber.

CHAPTER FOUR

IMPROVING THE WOODLOT

For most Canadian woodlot owners, the only management treatment the forest receives is the harvesting operation. The rest of the management is simply left to nature. Trees, like other crops, respond favourably to many different types of inputs. Tending the stand improves the growth, quality and value of the trees and enhances the aesthetic and recreational benefits derived from the woodlot.

Silviculture is Latin for "forest cultivation" and refers to the regulation of forest establishment, composition, growth and harvest of managed forests. An established forest can be altered to meet management objectives by a series of forest treatments called "stand tending." Stand tending continues throughout the life of the forest and falls into six categories: 1) weed control, 2) fertilization, 3) thinning, 4) pruning, 5) fire control, 6) pest control and 7) forest access.

Weed Control

"Weed" refers to any unwanted grass, herb or shrub. Weeds usually become a problem after harvest as canopy removal allows light to reach the forest floor and promotes growth of low-lying vegetation. Abundant weed growth can severely impede tree seed germination and inhibit growth of newly planted tree seedlings.

Weeds can be controlled in a number of ways. An easy, cost-effective method is by altering the type of harvest to a selection cut: the trees left standing block sunlight and take up water and nutrients that would otherwise go to weed growth. The effectiveness of this method diminishes proportionately with the percentage of trees harvested. Selection cutting is not recommended where regeneration of species that require full sunlight, such as pine and poplar, is desired.

Herbicides can be applied to small areas using a backpack sprayer.

Weeds may also be controlled by grazing livestock such as cattle or sheep. Livestock consume the weeds and then return much of the nutrients back to the newly established tree seedlings in the form of manure. Livestock can be extremely efficient and cost effective, especially for a type of thinning operation called "conifer release" (SEE CHAPTER 8).

Weeds can also be controlled by manually cutting the weeds with brush saws. This method is most effective if cutting is completed before the weeds set seed. It can be dangerous and expensive, and certain brush species such as willow can sprout many shoots from a single cut stem, thus compounding the brush control problem instead of controlling it. Manual brush management is a viable alternative in very small or very valuable stands of trees.

Chemical brush control can also be effective, but may also be expensive and dangerous. Small areas can be sprayed with a backpack sprayer. Larger areas, because of the undulating terrain and deadfall scattered throughout the brush area, require chemical application from the air by airplane or helicopter. Chemicals can leach into the groundwater, drifting into adjacent vegetation and possibly accidentally being ingested by animals or humans. Before initiating a spraying program, check with the local provincial forestry official for restrictions and necessary permits.

Fertilization

Nutrient addition is generally not cost effective for most rural woodlots, but may be justified for some high value stands within 10 years of harvest. Nutrient application, however, has its greatest potential in urban woodlots and land adjacent to centres of population.

Every day the urban environment generates massive quantities of liquid waste, which is expensive to treat and dispose of. By irrigating the urban woodlot, trees naturally and economically purify the nutrient-rich waste water, while providing an aesthetically pleasing forest and reducing the amount of water flowing through municipal water treatment plants (SEE CHAPTER 3).

Tree fertilization is also a byproduct of silvipasture and using animals for weed control. Many nutrients removed by the grazing and browsing of livestock are returned to the forest in the form of manure and urine.

Thinning

Thinning is the process of reducing stand stem density to allow the remaining trees to maximize fibre production by allowing unrestricted uptake of water, light and nutrients and to avoid mechanical abrasion of other trees. Thinning has two advantages: it enables the remaining trees to acquire the growth potential of trees that otherwise would have died from overcrowding, and it allows the production of a merchantable forest in a shorter time.

Thinning treatments can be applied throughout the life of the forest. The first thinning is called "juvenile spacing" and is used to prevent overcrowding of the very young stems. Because these small stems are often cut and discarded, juvenile spacing is called "pre-commercial thinning." Juvenile spacing treatment may have economic potential on smaller woodlots. Instead of cutting and killing excess saplings, they may be dug up and sold, or transplanted to an understocked area.

Juvenile spacing is usually carried out during the spring and early summer when the trees are still visible. Lush vegetative growth in late summer and snow in the winter can hide many saplings.

Juvenile spacing programs vary according to species, site conditions and management objectives. The ballpark thinning recommendation is to maintain juvenile spacing at about 3 to 4 m (10 to 15 feet) between trees, which will result in 700 to 1,000 evenly spaced trees per hectare As a rule, poplar trees are self-thinning and do not require additional chemical or mechanical treatments to reduce population density.

A special type of juvenile thinning similar to weed control is called "conifer release." Young white spruce and balsam fir will not exhibit rapid growth until the fourth or fifth year, usually when the tree is about knee height. If competing vegetation has formed a canopy, the spruce and fir will continue to grow slowly because of the absence of adequate sunlight. Once the canopy is removed, the spruce and fir will release, that is, exhibit tremendous upward and outward growth.

As the forest ages, each thinning operation that produces trees with sufficient size to be commercially valuable is called a "commercial thinning." Often the commercial thinning produces trees that only have marginal value as fenceposts, fence rails or firewood.

If the trees harvested from the commercial thinning are to be used as fence rails, the best time to thin is in late May and early June when the cambium is actively growing. The bark is very easy to peel during this period and often will be removed during the felling and skidding operation.

Target tree densities for mature trees at harvest vary from about 300/hectare (120/acre) for white spruce, balsam fir and jack pine, to 400/hectare (160/acre) for tamarack, larch and black spruce, and 500/hectare (200/acre) for lodgepole pine.

More care must be taken when thinning mature stands than juvenile stands because the greater surface area of the canopy makes some species to susceptible to wind and snow. Dense stands of shallow-rooted trees such as white spruce must be lightly thinned in several intervals separated by three to four years to develop wind firmness. Trees with large taproots such as the lodgepole and jack pines are more resistant to wind damage and therefore may be thinned in one operation. Areas with heavy winter snowfall should also be thinned in intervals to allow individual trees to develop thicker trunks to be able to withstand heavier snow loads.

Managing a woodlot can double stand productivity compared to an unmanaged forest because of the ability to closely monitor and quickly react to unexpected problems. Trees can be remain fully stocked as they grow and thinning treatments can be carried out only as necessary. Sanitation thinning, to remove insect-infested or diseased stems as problems develop, and improvement thinning, to remove any undesirable deciduous or coniferous stems, can be continuous processes throughout the life of the woodlot.

Pruning

Although thinning will increase the quantity of wood removed from a woodlot, pruning will increase the quality of wood in the woodlot. Pruning treatments are also beneficial for a number of other reasons including access improvement and fire control.

Wood strength and value is inversely proportional to the amount of knots present. Knots are caused by wood growing around branches and branch stubs and reduce the strength and corresponding value of the wood. By removing the branches early in a tree's growth, the sapwood is able to grow in a continuous band around the tree trunk without the interruption of branches. This knot-free wood is called "clear wood" and commands a much higher price especially in a plywood application. Many Asian countries will pay a premium for clear wood for the production of beams used in home construction. Pruning is unnecessary for trees slated for pulping, particle board or chipboard feedstock.

Pruning can be started when the tree is as small as 2 m (6.6 feet) tall. Pruning young trees has several advantages: the tree will yield more clear wood at harvest, pruning is much quicker and easier when the branches are small and hand-held pruning shears can be used, and ground fires cannot use lower branches as a ladder to destroy the crown.

Pruning the bottom third of this juvenile spruce (right) will prevent a ground fire from becoming a crown fire and provide additional clear wood when this tree is harvested.

A pruning saw or hand shears (above) can be used for smaller lower branches.

A bow saw (left) or long-handled pruning saw should be used for higher branches.

On taller trees, branches should be pruned to a height of 3 m (10 feet). This pruning, often called the "first lift," will ensure 2.5 m (8 feet) of clear lumber (standard height for plywood, two by fours, etc.) plus .5 m (2 feet) for the stump height. Only a third of the total height of the tree should be pruned for most tree species, a half of the total height for pine.

The best time to prune is in the winter when the trees are dormant and risks of insect attack and disease infestation are low. Winter also offers several other advantages. If trees are pruned before Christmas, the branches can be made into ornamental wreaths and sold. If the branches are not used, the snow cover and the cold temperatures allows for their safe burning right in the woodlot.

All branches dead and alive should be removed from the intended pruning area on the tree. Branches should be pruned flush with the bark exterior. Broken branches and pruning butts take many years for the tree to heal over and produce clear wood.

A chainsaw may be used for pruning, but take great care to avoid damaging the bark on the trunk. Also, for safety reasons the chainsaw should not be use for pruning branches above waist height For higher branches, use a pruning saw with an extended handle.

Access improvement is a common reason for tree pruning. Interlocking branches make travel very difficult for humans and animals. Height of pruning should be to the level where the branches do not interfere with any part of the body of a walking person especially if branches can poke the eyes. If access is for people on horseback, increase the height of pruning accordingly.

Fire Control

Fire hazard in the woodlot may be reduced in a number ways including pruning, reduction of fuel sources, correct timing of forest treatments and construction of recreational facilities in areas with low fire potential.

A dense stand of pine prevents forest access.

Often ground fires pass right through a forested area without causing a great deal of damage if the fire is not allowed access into the crown. Crown fires begin when ground fires use the lower branches to climb into the crown. By removing the lower 1.5 m (5 feet) of branches, the fire is unable to gain access to higher levels of the tree.

The major requirement for a fire to burn is a fuel source. Forest fire fuels essentially fall into two types: 1) slow, prolonged burning fuels such as standing timber, fallen trees, snags and sound stumps, and 2) easily ignitable, fast-burning slash and brush cuttings. Type of fuel affects both intensity and duration of a fire and usually both types can be found in all unmanaged stands of timber.

One way to reduce fuel sources to non-hazardous levels is by removal. Obviously it is not usually practical or even necessary to eliminate every last stick of combustible material on the forest floor, but areas with a large amount of potential fuels such as blowdown areas or slash piles can be removed by burning in the winter when conditions are not hazardous to the rest of the forest.

Forest fires often start as ground fires in dry grass by a carelessly thrown cigarette or a vehicle with a hot tailpipe. Grass and small brush can be removed from the forest area before it becomes a hazard by the use of herbicides, mechanical removal or grazing sheep and other types of livestock.

Timing of forest treatments is extremely important in reducing fire hazard. Restricting activities such as harvesting and pruning to the late fall or early winter when the weather is cool and snow depth is not yet a problem will all but eliminate accidental fires. All burning of brush piles and residual slash should be carried out in the woodlot only when the ground and surrounding trees are protected by a layer of snow.

Operating machinery equipped with fire prevention equipment such as chainsaws with spark arresters will eliminate potential ignition sources. Fire extinguishers should be in all vehicles present in the forest to suppress fires before they get out of control. Fire can be averted by performing some simple precautionary measures such as not smoking while refuelling, never refuelling a hot chainsaw and moving away from the refuelling area before starting the chainsaw.

Proper placement and construction of recreational facilities, especially firepits and barbecues, can also be critical in reducing fire hazard. Firepits should be located in areas of sand or mineral soils away from organic "peaty" soils. The firepit should be lined with an outside layer of sand and an inside layer of rock or other non-burnable material. A steel oil drum cut in half and buried in sand makes an excellent firepit.

Ideally, firepits should be located near a source of water such as a creek, lake, pond or dugout in the event of an accidental fire. Firewood and other flammable material such as lighter fluid should be stored well away from the firepit. Place a low steel fence or expanded metal grate around the fire to prevent inebriated people and small children from falling in.

Fire prevention is the most important aspect of forest fire protection, but a fire suppression strategy is necessary in the event of a fire. Early detection and functioning firefighting equipment are essential in controlling fires once started. Pumps, hoses and spraying nozzles are

With a few simple precautions this woodlot fire could have been avoided.

A properly constructed firepit (top) allows for a worry-free hotdog roast.

This portable water pump (bottom) can pump water from a creek or pond or it can be attached to a portable water tank.

effective if a water source is nearby. Portable firefighting equipment can be built very cheaply with parts salvaged from agricultural spraying implements.

Well-maintained woodlot access in the form of roads and trails facilitate regular forest monitoring. Fire hazards can be recognized and allow firefighting equipment and personnel to reach a blaze in the shortest possible time.

Pest Control

Each year diseases and insect infestation cause millions of dollars of damage to public forests. Losses occur in the form reduced growth, decreased wood quality or outright killing of the trees. Associated injuries include lengthened harvest intervals, change of species composition, and degraded water and soil. Pest outbreaks can usually be prevented by maintaining a blend of ages and species, performing silvicultural treatments and providing plenty of forest access.

Perhaps the simplest and most effective method of pest control is to maintain an uneven aged mixed-wood stand, or a mosaic of small plots of same age, same species. Such a configuration breaks up the woodlot and prevents rapid simultaneous outbreaks through the entire woodlot because most diseases and insects are species specific.

Forest pests, in general, are much easier to prevent than to control. As with humans, healthy trees are more resistant to various maladies than trees that are stressed. Healthy trees come from initiating proper forest treatments. Successful pest management is often an unnoticed byproduct of successful woodlot management.

Ample forest access at regular intervals through the woodlot is another significant element in a pest control program. By allowing for regular and frequent forest monitoring, early detection of a pest problem and timely application of the appropriate remedy or removal of the infected trees can prevent further outbreak.

The fungus on the outside of this aspen trunk indicates heart rot. The background shows that the infestation has spread to all the trees in this even aged aspen stand.

Forest Access

The development of roads and trails is important in all aspects of woodlot management including forest monitoring, logging and maintenance. Access adds dramatically to the recreational value of the woodlot whether for walking in the summer or skiing in the winter. Access is also necessary for the movement of livestock and wildlife.

For simplicity, forest access will only be separated into two categories: roads and trails. A road is defined here as an elevated all-weather surface requiring heavy equipment and megabucks for construction. The vast majority of private woodlots are served by an existing network of municipal roads. Contact the local forestry office for guidelines on road construction.

Trails differ from all-weather roads in the quality and expense of construction. Trails generally only have surface obstacles removed, are not ditched and the trail surface is not improved with gravel or pavement. Trails usually do not suffer the severe erosional problems and are aesthetically more pleasing than all-weather roads because much of the grass and low-lying vegetation is left intact.

The easiest and perhaps visually most pleasing way of trail construction is to follow the path of least resistance. By following elevation contours and avoiding as many trees as possible, many trails can be built quickly and inexpensively.

Trail construction can usually completed with the minimum of equipment. In a mature forest where a complete canopy prevents full sunlight from reaching the forest floor, usually a handsaw and a pair of pruning shears is all that is necessary to remove the underlying vegetation and overhanging branches. In juvenile stands, a chainsaw and brush saw may be required. In new plantings, seedlings can be positioned for the insertion of future trails.

Trails should be wide enough to accommodate forestry and trail maintenance equipment. Avoid sharp corners and switchbacks if possible, to facilitate ease of movement for long loads and to make skiing easier and safer.

This bush (left) was turned into a 2.5 m (8 feet) wide trail (right). By choosing the path of least resistance, no trees with a diameter at breast height (dbh) of greater than 10 cm (4 inches) were removed and clearing took less than one hour per 100 m (330 feet) of trail.

Sharp trail bends should be avoided because moving logs will create "skid scars" on standing trees. The skid scar at the base of the tree (right) provides an entry for parasites, insects and diseases.

Constructing a trail along a side slope is also easy and cheap with a backhoe, bobcat or tractor with a front-end loader. Dirt from above the trail (cut slope) is placed adjacent to the trail on the down slope (fill slope).

The best time to construct a side slope is in early autumn when the chance of massive erosion caused by rain from thunderstorms is diminished. Fall rye or other winter annuals underseeded to lawn grass can be established late in the autumn and still provide sufficient root growth to prevent erosion on the cut and fill slopes from the spring meltwater.

Stream crossings can be constructed in a number of ways including fording, bridge building or using culverts. Building a ford involves the placing sufficient rock on the streambed and banks to support the weight of vehicles without causing extensive erosion. Water can still flow over and through the rock material. Fords are usually only practical on small streams with shallow banks and are not recommended for walking trails unless wet feet are desired.

Bridges come in a variety of shapes and sizes, some cheap and others expensive (SEE BELOW AND NEXT PAGE, for some bridges and their applications).

Installing culverts is often the most cost-effective method of stream crossing, especially when heavy logging equipment will be using the crossing. Either plastic or steel culverts can used with the size and number depending on stream flow. It is important to have enough culvert capacity to handle the maximum flow conditions that occur in the spring or after a heavy rain. A overflowing culvert may wash both the overlying material and the culvert itself down the stream.

For many small streams, plastic culverts can be positioned by hand. However, a backhoe or other equipment may be required for installation of large steel culverts. In either case, it is very important to line upstream culvert entrance with rock, sandbags, steel or any other heavy durable material to prevent erosion from the flowing water.

A snow bridge is suitable when a permanent bridge is not desired. The bridge is constructed by pushing snow onto the creek and pumping creek water on top to freeze and harden. The bridge can support large loads and disappears after spring thaw.

This permanent bridge is constructed from an old train flat deck and can support large loads.

This culvert with a cement cap is used where stream flow is generally low but is subject to extreme levels in the spring or during rainstorms. The cement cap allows water to flow over the culvert without washing it away.

This suspension bridge is aesthetically pleasing but is expensive and not suitable for heavy vehicle traffic.

These bridges are suitable for pedestrians and are inexpensive, but they might wash away.

This bridge is the simplest and cheapest bridge of all, but it is only suitable for agile pedestrians and might float away during periods of high water.

Steel culverts are often discarded from road construction and can be picked up for free.

After trail construction is complete, maintain and mow the trails periodically to prevent the grass from becoming a fire hazard and to make trail travel safer and more enjoyable. If you are using the trails for cross-country skiing in the winter, mow the trails before the first snow fall, as tall grass will promote early patchy melting.

Implementing woodlot improvement techniques will result in the landowner reaping a vast number of financial and aesthetic rewards. You will reap monetary benefits from a healthier, more productive forest, and also numerous intangible rewards such as the enjoyment of walking or skiing through a personal paradise.

A tractor-mounted rotary mower makes regular trail maintenance quick and easy.

Well-maintained trails can be safely used for family outings in the summer or for skiing in the winter.

Tree Trivia

The Athabasca Landing Trail, the Hudson's Bay Company's fur-trading road, when built in 1875, was 3 m wide (10 feet) and stretched over 160 km (100 miles) between Edmonton and the town of Athabasca. It had a total cost of construction of only $4,059.00.

Chapter Five

HARVESTING

To many people, cutting trees epitomizes forest destruction. On the contrary, the harvesting operation, far from being the conclusion, represents the most important component of rejuvenation on which all other forest treatments rely. Harvesting is a simulation of the natural catastrophic events such as wind and fire. Harvesting allows the removal of dead, diseased and unproductive trees and allows for new vigorous growth that can be used by humans and animals.

The correct harvest technique depends on the integration of a whole host of management goals. The most important consideration when selecting a type of harvest is the silvic requirements of species. Other factors such as terrain, recreation potential, wildlife habitat and economic potential of the standing and future forests must be seriously considered before even one tree falls. A single harvest mistake such as improper haul road construction may return to haunt a woodlot manager for many years.

Woodlot harvesting strategies are as varied as the person who manages them. Some harvest operations such as beaver logging (SEE CHAPTER 8) rely entirely on animals for felling and delimbing operations. More traditional harvesting strategies use modern power equipment and the most up-to-date silvicultural techniques. Forest harvest usually falls into one of two categories: 1) clearcut (even aged management), where all trees are harvested in a given cut block, or 2) selection cut (uneven aged management), where only some of the trees are harvested from a given cut block.

Clearcuts

Clearcutting, contrary to many popular misconceptions, is an excellent harvest strategy in many situations. Clearcutting is the harvest option of choice in even aged stands, overmature stands, stands to be replanted to shade intolerant or pioneer species, and for development of wildlife habitat of popular game species. Clearcutting is beneficial in certain salvage cuts for removing large areas of trees that are diseased, burned or blown down, and also tends to be the safest, most economical type of harvest when harvesting with large machinery.

Canada's largest forest region, the boreal mixed-woods, has developed species that depend on short-term periodic catastrophe such forest fires to rejuvenate old stands. Clearcutting is the harvest operation that most closely mimics the natural succession. Intense provincial forest fire control, which began in the late 1950s, has resulted in many overmature, even aged stands, especially in agricultural areas. The overmature trees tend to be less vigorous, less productive and more susceptible to disease, insects and parasites.

Unmanaged woodlots often contain one or more overmature stands. Harvesting these stands while the wood is still sound and valuable makes good economic sense and can provide cash flow for proper management of the rest of the woodlot. Trees allowed to die of natural causes cannot be used in certain commercial applications, thus limiting selling options. Standing dead trees also pose a hazard to woodlot workers and recreational users.

Clearcuts are the harvest method of choice when large equipment such as this feller-buncher is being used.

Clearcutting is also the best harvest method when regeneration of shade intolerant pioneer tree species such as poplar, birch, tamarack, jack pine and lodgepole pine is desired. Complete canopy removal allows full sunlight to reach the forest floor. Warm soil promotes asexual vegetative reproduction (suckering) of poplar and temperatures high enough to melt pine cone resin to facilitate seed release from the pine cones left in the logging slash. Full sunlight is also required for maximum growth for the emerging seedlings.

Wildlife can also benefit from clearcuts. The canopy removal results in the growth of grasses, shrubs and saplings while the adjacent standing forest provides "edge." Forest edge allows primary consumers such as deer, elk, moose, grouse and rabbits to graze or forage on the succulent new growth with protective cover close by. Predators such as hawks and foxes are also attracted to the woodlot edge because of the abundance of prey.

Natural disasters such as forest fires, wind storms or insect attacks can kill large areas of productive forest. It is usually necessary to harvest these stands within two years of the catastrophe otherwise natural degradation of the wood results in dramatic reduction of industrial applications and corresponding value. Depending on the cause of the kill, unharvested downed trees soon become insect-breeding centres or fire hazards and should be clearcut and removed before the remaining standing forest is threatened.

Clearcuts also tend to be the safest and most economic method of harvesting large tracts of land. Productivity is maximized by use of heavy equipment and the per hectare cost of fixed infrastructure such as road and trail construction and maintenance is minimized. Reforestation costs may also be reduced when planting seedlings and a clearcut is the only option when regeneration of pioneer species is desired.

Clearcutting a forest also has a number of potential drawbacks. Often such harvests are regarded as unsightly by neighbours or visitors. Improper cutting- or road-building techniques on hillsides may result in excessive soil erosion. Discretion is also required when clearcutting forests located on stream or lake margins (riparian zones, SEE PAGE 118) to prevent degradation of water quality caused by increased water temperatures and silt accumulations.

Selection Cuts

A selection cut is the individual removal of certain designated trees in a forest. Selection cuts are most often used to maintain an uneven aged stand and in stands where regeneration of shade tolerant species such as white spruce or balsam fir is desired. There are essentially five types of selection cuts: 1) thinning cut, 2) seed tree cut, 3) high grade cut, 4) shelterwood cut and 5) salvage cut.

Thinning Cut

A thinning cut is performed in stands where overcrowding prohibits obtaining maximum value of the stand. Thinning cuts can be divided into two categories: 1) precommercial thinning, and 2) commercial thinning. Precommercial thinning is usually performed in juvenile forests or plantation stands where the undesirable trees are too small to have market value and they are discarded. Precommercial thinning is described in more detail in Chapter 4.

A commercial thinning cut typically occurs in a more mature forest where the trees designated to be removed are of a merchantable size. Usually the trees selected for removal are of an undesirable species or are defective in some way. Individuals exhibiting poor growth, deformed trunk or disease symptoms are extracted and often marketed as firewood or as feedstock to pulping operations. Small diameter logs can be safely treated with chemicals and used as fenceposts or rails.

This large spruce was left as a seed tree to stimulate natural regeneration in this cut block.

Seed Tree Cut

Seed tree cuts result from the continuous commercial thinning of a stand where only a few of the largest, most desirable trees are left to produce seed for the next generation of trees. After the new crop of trees has become established, the seed trees may then be harvested or they may be left if

This spruce forest had all the superior trees harvested, as the stump in the foreground indicates. The crooked, slow growing trees were left to produce seeds for regeneration. (Note the aspen regeneration.)

harvest would require excessive damage to the new growing forest.

Seed trees can be beneficial in a number of ways. Natural regeneration tends to be cheaper than initiating a planting program. By removing less desirable trees, the genetic potential of the seed source for the next rotation is likely to be high. Cut blocks can be much larger than naturally seeded clearcuts because the seed source comes from trees scattered throughout the cut area and not just the adjacent forest. Seed trees can also be used specifically for cone harvest for greenhouse seedling production.

There are several disadvantages to the seed tree cuts. First, natural regeneration can be unreliable and may result in overstocked zones at the base and downwind of the seed trees and understocked zones in other areas. Second, by leaving the largest, most valuable trees to be harvested last, the initial short-term cash flow is reduced. Third, harvesting the seed trees can result in a large amount of damage to the newly established seedlings by the harvest equipment. Last, forest regeneration using seed trees works best with shade tolerant species.

High Grade Cut

A high grade cut (or high grading) is the practice of harvesting only the biggest and best trees, while leaving the less desirable trees standing. The main advantage of a high grade cut is that short-term cash flow is maximized.

High grading has most of the disadvantages of a seed tree cut with none of the benefits. The trees left to produce seed for the successive generations tend to be genetically inferior and consequently subsequent rotations will be less productive. Because much of the remaining forest is of low value, further management is no longer cost effective.

Woodlot managers must use extra caution when selling standing timber to buyers who offer to selective cut the woodlot for a fixed price. Unscrupulous operators will often attempt to high grade the forest to maximize profit. Such contracts should clearly outline for criteria for timber to be cut.

Shelterwood Cut

A shelterwood cut is the method of leaving certain trees standing for the protection of subsequent generations of trees. Shelterwood cuts can comprise small clearcut strips or the removal of a nurse crop.

Shelterwood strips are created by cutting narrow, long bands of trees through the woodlot. The width of the bands is determined by the quantity of shelter and the sunlight required for the next rotation of trees. The standing trees provide less shade and less wind protection when the opening is wider.

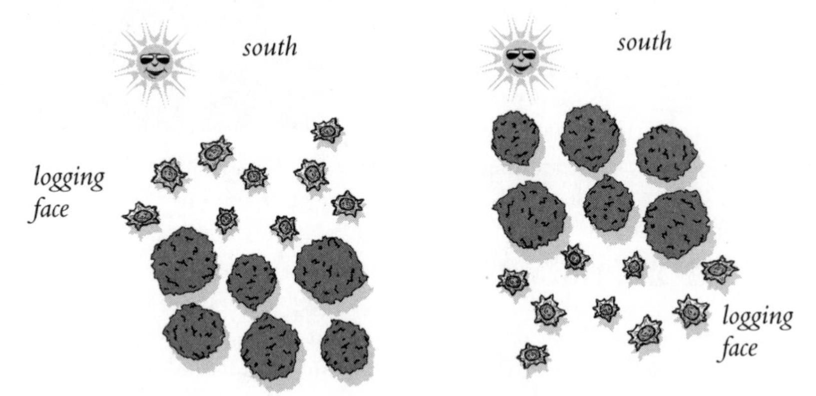

Two types of shelterwood cuts: for shade intolerant trees (left), and for shade tolerant trees (right).

Orientation of the successive shelterwood strips also depends on the species to be regenerated. If shelterwood strips run east and west, then each additional strip should be cut on the north side of the first strip for shade intolerant tree and on the south side for shade tolerant trees.

The length of the shelterwood cut is most often determined by woodlot size. On large forests, however, the length and orientation of the shelterwood strip must be adjusted to prevent strong winds from funnelling between the remaining trees and causing blowdown. Blowdown can be avoided by cutting the strips in the shape of a herringbone.

Other types of shelterwood forest involve using a nurse crop. The nurse crop or overstorey is usually a shade intolerant pioneer species such as poplar or pine. The understorey generally comprises shade tolerant species such as white spruce or balsam fir. These nurse tree/understorey tree associations occur in nature as part of the normal succession sequence following a natural catastrophe such as fire.

The apical bud of this white spruce is in continual contact (leader whip) with this mature aspen (left). A damaged leader allows competition from lateral branches, resulting in a multiple leader (above).

As the understorey grows, the apical bud (leader) contacts the branches of the nurse crop. Often, especially after violent storms, the apical bud becomes damaged by this contact in a process called "leader whip." The lateral buds then compete for dominance, which may lead to the formation of multiple stems from a single tree. Removal of the nurse crop is recommended when leader whip is observable throughout the forest.

The advantage of the shelterwood cut is that the forest is systematically cut and replanted in a number of separate events. Cash flow can be regulated, as can forest improvement practices. A variety of timber sizes and ages are available for different types of wildlife habitat.

The shelterwood nurse crop reduces rotation times on two separate species of trees. It allows the woodlot manager to reforest the stand to a second generation before the first stand is cut, thus optimizing labour and machinery. Mixed-wood nurse crops may have greater productivity because of different nutrient utilization by the different species. The mixed-wood forest also offers protection against species-specific outbreaks of disease and insects.

The disadvantage to managing a shelterwood system is the difficulty of mechanical harvest. Large mechanized equipment tends to damage the remaining trees. Shelterwood harvest is more suited to smaller specialized equipment or horse logging (SEE CHAPTER 8).

Salvage Cut

The salvage cut can include aspects of all the harvesting techniques mentioned above. Salvage cuts are the harvest technique of choice in urban forests and in very small woodlots. Salvage cuts involve removing any dead or dying tree that has succumbed to wind, fire, disease or other maladies. Trees that touch power lines, pose a safety hazard, or interfere with TV or radio reception can be also included in this class.

Salvage cuts can be clearcut, especially in forests that have been burnt or blown down, but typically involve only a few trees. Beaver logging (SEE CHAPTER 8) is also another good example of salvage logging.

In some municipalities, rights-of-way for roads and power lines are cleared and the fallen trees are left to rot or are pushed in a pile to be burned. In urban areas, trees cleared for new subdivisions have the same fate. Permission to remove the wood is often willingly given because it saves the contractor the expense of disposing of the "waste" wood.

Many salvage opportunities exist right in the heart of large urban centres. The outbreak of dutch elm disease (DED) in Canada has killed thousands of elm trees lining the boulevards of many towns and cities. DED does not harm the wood and hence its value as lumber is not re-

This blowdown should be salvaged or many types of insect pests will make their home in the decaying wood.

duced. Using dead boulevard trees for saw logs does have a disadvantage: the wood often contains nails and tacks from posters, etc., which can damage a saw blade.

The list for urban timber salvage opportunities is enormous because the original intention for tree planting was for ornamental value only. Once the tree has died, it is valued as worthless. It is cut and may end up in a landfill. A smart urban lumberjack can make money two ways: by removing the tree and by processing it into lumber or firewood.

Harvest of burned or blown down areas must be completed within two years, after which the soundness of the wood deteriorates rapidly.

Harvest Contracts

For many woodlot managers to perform the harvest operation themselves is not feasible. Time, money and labour constraints make it necessary to hire professional forest contractors to conduct the harvest operation for you.

Hiring a harvest contractor can be beneficial in many ways. A qualified contractor should have the experience, specialized equipment and forest "savvy" to execute the harvest operation with maximum efficiency and minimum damage to the environment. The contractor is responsible for all safety and Workers' Compensation Board (WCB) requirements, machinery repair and downtime in the event of poor weather conditions.

Check out the contractor's credentials *before* commissioning work. Ask for references and talk to the people for whom the contractor has worked. Walk through some of the areas the contractor has harvested. Make sure that the contractor has an account with the WCB, because landowners can be liable for all work-related injuries that occur on their property. Usually the best large mills provide a certified and bonded contractor to do the work.

Once a contractor has been selected, it is important to draw up a *written* contract. The purpose of the contract is to outline all the obligations of the woodlot manager and the contractor. Details about the volume, species and size of trees to be harvested and other conditions such as cleanup, reforestation, etc., must be specified in the contract. If problems arise in the contract it is best to retain the services of a forestry or a legal expert. What appears to be a small concern now may take much time and money later battling out in court.

The disadvantages to hiring a contractor include lower return per unit volume of wood sold, incomplete control over woodlot practices, idle equipment and labour, and the intangible benefits such as exercise and pride in doing a job well done.

Falling and Hauling Equipment

Once it is determined that the harvest operation is to be done inhouse, it is essential to gather all the required equipment and make it "field ready." The harvesting process requires essentially two elements: 1) a method of cutting and delimbing the trees, and 2) a method of hauling the trees to a point of sale or processing.

For small woodlots or urban forests, the cutting and delimbing operation is usually done with a chainsaw, handsaw or axe, and hauling is usually in the back of a pickup. With a larger woodlot, hand work becomes impractical and it is necessary to use progressively more mechanized equipment.

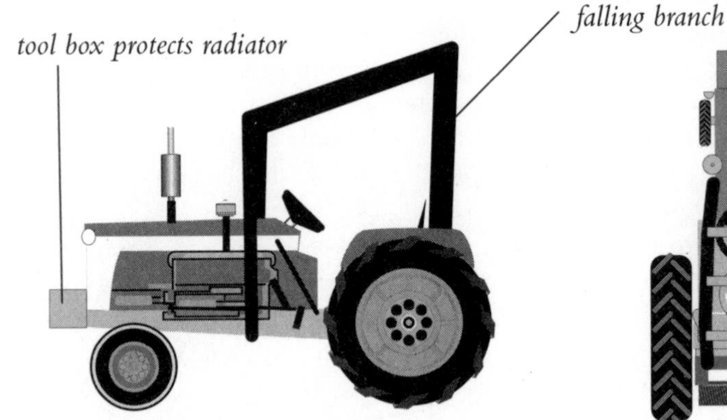

Refurbishing an older tractor to do woodlot work is inexpensive and easy to accomplish.

This small tractor is equipped with a three-point hitch round bale fork. The logs are skidded parallel (left) with the log deck and then stacked with the bale forks (right).

Unless very large acres are to be harvested, the chainsaw is usually the cheapest and easiest method of falling and delimbing trees. The type of chainsaw is a personal choice, and the advice of a professional faller will help an inexperienced lumberjack select the make and model suitable for a particular application.

It must be remembered that falling trees by hand is a **VERY DANGEROUS** operation. No would-be faller should ever venture out into the forest without prior training and proper safety equipment. Chainsaw safety is discussed in more detail later in this chapter.

Farm woodlots have equipment available which can easily be converted to a timber harvesting application. A few simple modifications to the farm tractor enable it to cut, transfer and load logs. Log trailers can be cheaply manufactured in most farm shops so the timber can be quickly and cheaply moved from the forest landing to the secondary processing facility.

The most important aspect of modifying a farm tractor for work in the bush is to ensure complete protection of the machine and operator. Minimum tractor safety features include a rollover protection structure (ROPS), a well-designed seat and safety belt, slip-safe steps, tow hook and power take-off shield. In addition, the ROPS should have a roof and expanded metal mesh screens on both the rear and the side. A bracket to hold a first-aid kit and fire extinguisher in the tractor cab is also safety must. Although most modern tractors come with cabs equipped with many of these safety features, older models require additional structures installed.

Protection of the tractor is another important consideration before the tractor is used in the forest. Expanded metal screens should be positioned to protect the radiator, headlights and engine sides. Belly pans to shield the underside of the engine and tire valve stem protection is absolutely essential to prevent tractor downtime in the forest. Tire chains, although not absolutely necessary in dry level conditions, will go a long way to increasing productivity when snow or mud is present. Dual wheels should never be used when using the tractor in the forest as logging debris will wedge between the wheels and damage the tires.

Many types of farm tractors are suitable for work in the woodlot, although tractors with four-wheel drive or front-wheel assist have superior traction and pulling power, especially in deep snow. Tractors mounted with front-end loaders are a real advantage when stacking logs at the landing or loading a log wagon. Tractors with a three-point hitch can skid large logs by lifting the large butt ends using chains mounted on a spreader bar or bale forks.

Although the tractor is the basic requirement for log transfer, additional tractor-mounted forestry equipment is available. Falling equipment, winches and cranes are attachments that will increase productivity of the harvest operation. Tractor- or trailer-mounted grapple loaders are also available for the loading operation, but in most cases a front-end loader with a round bale grapple attachment is sufficient.

A PTO-driven winch attached to the three-point hitch is able to pull logs up to 50 m (164 feet) away from the trail, reducing the amount of trail necessary to harvest a given area. When selecting a winch, do not choose one that can pull more than the tractor weighs, otherwise the winch may end up dragging the tractor backward if a log jams.

Just about any farm tractor can be converted for moving woodlot timber. One tractor (above left) is equipped with a high lead boom and a PTO-driven winch, and the other (right) can move even the biggest logs with ease with its front-end loader and bale grapple.

A hydraulic-driven winch mounted on this small tractor can save time and money when hauling out logs from the bush (right).

Most managers of farm woodlots simply convert hay wagons to transport logs (below left). Small log/lumber hauling trailers can be built very cheaply using old drill pipe and salvaged running gear (below right).

Log-hauling equipment can be made by adding vertical bunk ends to flatbed hay trailers equipped with single or bogey wheels. Other types of trailers to fit behind a pickup truck can be made from tubular steel. Many farm equipment dealerships carry multiple axle log-hauling wagons. Trailers perform better than wagons because the trailer tongue pressure helps increase traction, and trailers are more manoeuvrable and not as susceptible to jack-knifing when backing out of a tight spot.

The most important consideration when building or buying a wagon is the weight-to-axle ratio if the logs are to be hauled down municipal or provincial highways. Always check with local transportation authorities for information on log-hauling permits and regulations before venturing on improved roads with a load of logs. It is also necessary to obtain a Forest Products Hauling Record (FPHR) if transporting logs on municipal or provincial roads. The FPHR is necessary to ensure that loggers are not illegally harvesting private or crown land.

Discussing your harvest plans with the local forestry or conservation officers before falling the first tree is usually a wise idea.

The Harvest Operation

It is not enough to obtain the proper equipment and head out into the bush and start harvesting. It is very important to have a well thought-out harvest plan in place before firing up the chainsaws to minimize the likelihood of accidents and maximize equipment and labour productivity.

The harvest plan should be organized so that all short- and long-term goals for the woodlot are taken into account. Considerations such as economic viability and environmental impact must be clearly thought out before the first tree has fallen. It may be good idea to detail harvest intentions to the local provincial forest official or a forest consultant for suggestions and comments.

Unless the logs are for personal use, it is imperative that price and quantity has been negotiated with the prospective buyer *before* harvest. Once the wood has been cut there is no way to restore logs back to living trees, if prices flatten out after harvest.

It is also necessary to acquire all the necessary permits and licences before initiation of harvest. When harvesting crown land, procuring a provincial wood harvest permit is a must. On private agricultural or urban land it may be necessary to obtain a municipal development permit for tree removal. A single phone call can make the difference between a successful harvest or an endless legal battle, so be sure and check it out.

If the wood is to be sold to a large pulping or sawing facility, it is often necessary to obtain a Workers' Compensation Board (WCB) policy for all personnel involved in the harvest operation.

The harvest plan should detail all the woodland operations including location and access to all harvesting areas, landings and skid trails. In a clearcut, the cut block boundaries must be clearly defined. In a selection cut, the fallers must be aware of the exact parameters of the trees to be harvested and the trees allowed to grow. If possible mark each tree to be cut.

The harvest plan should also include a centralized landing area. The landing acts as drop-off point for logs skidded from the forest, and a storage, sorting and loading depot for trucks hauling to secondary processing points. The landing should be located on a well-drained site accessible to an all-weather road.

No harvest plan would be complete without an Emergency Response Strategy (ERS) The ERS is a set of steps to follow in the event of an accident or natural disaster. The ERS should contain mitigation procedures for fire control and medical emergencies and detail location and operation of all the forms of safety equipment.

Safety

For many woodlot owners, the harvest represents the most important operation in the woodlot because of the revenue it generates. No amount of money, however, can return a corpse back to life or unparalyse a quadripalegic. In terms of WCB claims paid out, harvesting timber is one of the most dangerous occupations in Canada. Safety in the woodlot should be placed first and foremost in the minds of all workers in the woodlot regardless which treatment is being carried out.

First-time loggers may be able to identify their biggest safety hazard simply by looking in the mirror. Timber harvest is physically demanding and requires a large degree of cardiovascular conditioning and mental alertness. Using timber harvest as a weekend workout to "build up a sweat" may result in injury, disability or death. Timber harvest is one place where the saying "No pain, no gain" definitely does not apply.

People who suffer from dizzines, fainting spells, alcohol or drug abuse, in fact anything that compromises reaction time, should not consider timber harvest. The split-second hesitation caused by impaired alertness may be the difference between life and death.

Pre-Harvest Preparation

Before entering the forest to harvest trees you should…

Obtain harvest training from a certified instructor at a recognized forestry school. Chainsaw safety certification is required by many mills before timber will be accepted for delivery (top).
NOTE: Make sure every worker involved in the harvest operation is covered by a Workers' Compensation Board (WCB) policy. Many modern mills will refuse timber delivery unless a WCB policy can be produced.

Service all equipment and make sure it is functioning properly before heading out to the bush (centre).
NOTE: Make sure weather conditions are suitable for the timber harvest operation. High winds, or low visibility caused by heavy snow, rain or fog, can be deadly in a harvest operation.

Have each person involved in harvest operation wear the correct protective clothing: steel-toed boots, ballistic nylon falling chaps or pants, high visibility vest, heavy leather or ballistic nylon gloves, face shield and hard hat. The fanny pack contains a personal first-aid kit and a whistle (bottom).
NOTE: A fire extinguisher and large first-aid kit should be on site and accessible at all times. All personnel involved in the harvest operation should have first-aid training. All work using brush saws, chainsaws or other power equipment should be performed in groups of two or more people. If an accident occurs to one person, the other person can perform first aid or go for help.

Harvest Procedure

The proper procedure for falling trees is…

Plan:

1) Where to start so you can avoid falling trees on top of standing trees, resulting in leaners.

2) To eliminate as many natural hazards as possible: never fall a tree from a downslope position, never fall a leaner by falling other trees on top of it, and never fall a tree with broken branches in the crown (get assistance from the skidder operator).

3) To develop a "face" that takes advantage of the natural lean of the stand and is usually down from the prevailing wind (below).

Before cutting look up (above). Note overhead hazards such as leaners, broken branches, etc. From a distance a tree may look normal, but the top may be broken off and is suspended by lower branches. Such a tree is definitely a hazard and you should leave it for a tractor or skidder to push down.

logging face

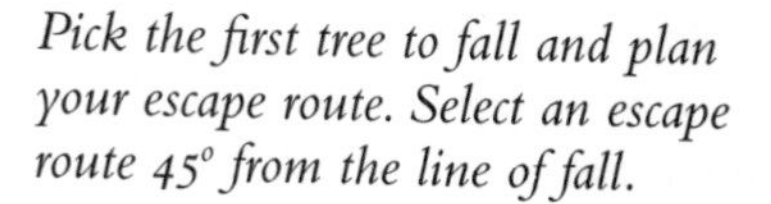

Pick the first tree to fall and plan your escape route. Select an escape route 45° from the line of fall.

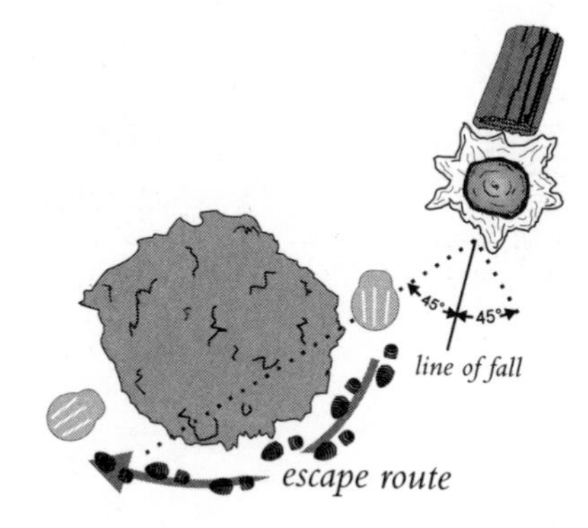

Remove all the brush and debris around the base of the tree and along the escape route (above). Check cutting surface to make sure that no obstacles such as barbed wire are present.

Notch the tree in the direction you want the tree to fall. Proper dimensions of notch (left) and actual notch (right).

Make side cuts if required. Side cuts are necessary when the diameter of the tree is longer than the length of your chainsaw bar.

Start the back cut. Proper dimensions of back cut (left) and actual back cut (right).

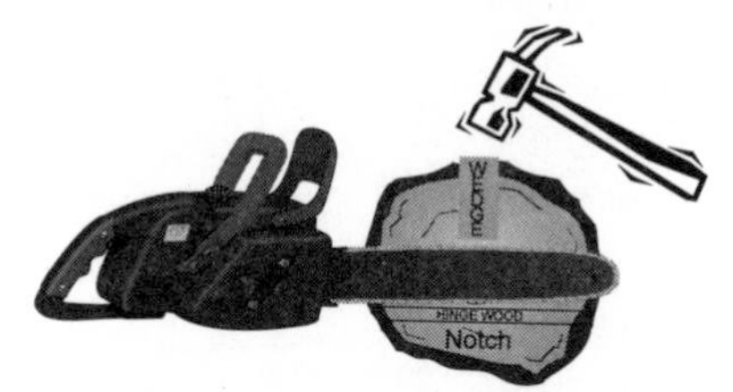

Start the wedge when the back cut is deep enough. Pound wedge in without hitting the saw blade. The wedge will prevent the tree from sitting back and jamming the saw blade and will direct the fall of the tree.

Complete the back cut, driving the wedge if the tree does not fall on its own. Retreat along the escape route as the tree begins to fall. Always keep an eye on the falling tree. Wait in a safe position until all overhead debris and loose or thrown branches have fallen (right). ***Remember*** *the "fatal 15," the 15 seconds that are the most hazardous. The fatal 15 lasts from when the tree first begins to fall until it is safely on the ground and all debris has fallen.*

Other Hazards

A leaner (left) should be pulled down from the butt end using a tractor or skidder (right).

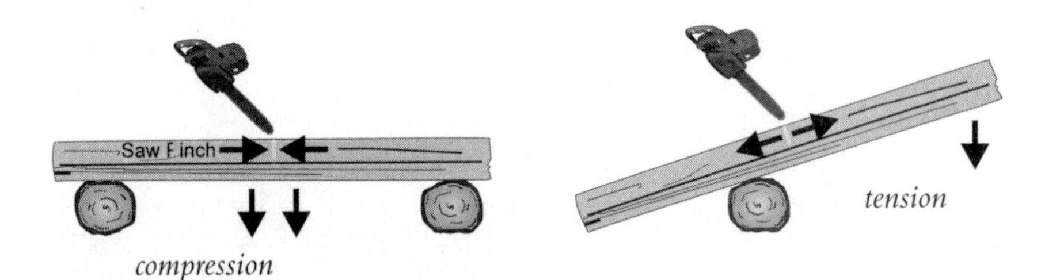

Assess the differential stresses on fallen timber. Trees under compression (left) may kick up, and trees under tension (right) may bind the saw.

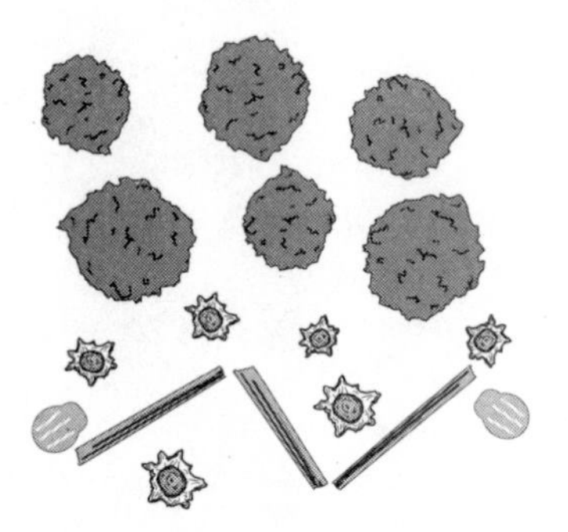

When two people are falling and delimbing on the same face, they should be a minimum of three tree lengths apart to prevent catapulting branches from injuring one another.

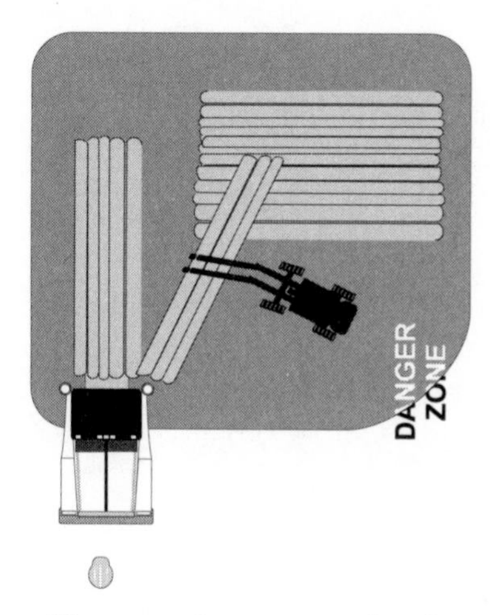

All personnel must recognize the extent of the danger zones when on-loading and off-loading logging trucks.

Tree Trivia

Falling trees and branches cause more than 60% of woodland injuries. Most of these accidents are attributed to improper falling techniques and failure to look up and assess potential dangers. 25% of woodland injuries result from chainsaws and brush saws and the remainder of accidents can be divided equally between vehicle accidents and improper loading.

Chapter six

REFORESTATION

Although many people associate reforestation with tree planting, in reality reforestation begins during the harvest operation. The amount and type of reforestation selected for a particular woodlot depends largely on the type and amount of harvesting that was carried out there previously. Your goals for the forest, woodlot size, finances and labour available, and type of trees to be restocked must all be considered in a regeneration plan.

There are four ways to regenerate a forest: 1) natural regeneration (nature handles everything), 2) pseudo-natural regeneration (site preparation), 3) artificial regeneration (planting trees), and 4) a combination of the above.

Remember that reforestation starts *before* the harvest of existing trees, not after. The harvest program selected will be the most important component of your reforestation plan, so it is very important to ensure that the type of harvest is compatible with the type of regeneration selected.

Natural Regeneration

The cheapest short-term management option is to allow natural regeneration to occur. Natural regeneration is the process by which the trees, stumps and roots that remain after harvest or natural disturbance provide the seeds or suckers to re-establish a forest with little or no site preparation from you.

Natural regeneration ideally suits reforestation of aspen and balsam poplar stands following clearcutting. Clearcutting creates conditions favourable for asexual vegetative reproduction (suckering). The suckering process is controlled by three major factors: apical dominance, soil temperature and available light.

Asexual reproduction by suckering in aspen is initiated when an imbalance occurs in two hormones (auxins and cytokinins). Cytokinins are found in the roots and promote suckering, while auxins occur in the crown and the trunk and inhibit suckering. When a aspen dies or is logged, the flow of auxins from the above-ground parts is interrupted, the cytokinins hormone becomes dominant and suckering is initiated.

High soil temperatures (above 15°C, or 59°F) degrade auxins and increase the cytokinin production, enhancing sucker production. A clearcut or fire-destroyed forest has its canopy removed, thus exposing the forest floor to direct sunlight and resulting in increased soil temperatures. Cool soil conditions, such as when the forest floor is shaded by unharvested trees, when it is covered by large amounts of organic matter (duff), or when the forest occurs in extreme northern areas, are not conducive to natural regeneration.

Poplar trees can also regenerate by seed. Mature poplars can produce as many as 1.5 million seeds per year which can be carried by the wind over 100 km (62 miles). Seed regeneration can be an important method of regeneration in cultivated fields, logged areas with site preparation and in northern areas where cool soil temperatures inhibit sucker production.

Natural conifer regeneration is confined to reproduction from seeds only. Therefore it is necessary to have a seed source near the proposed regeneration area. Pine and tamarack are pioneer species that require full sunlight and respond best to clearcut sites that have seed trees remaining on the perimeter. The ideal size of the clearcut will vary according to how much area is available for cutting, but for optimum results diameter of cut areas should not exceed 200 km (124 miles).

Natural regeneration of species present in an existing forest can also be consistently successful in selection cut and seed tree climax forests of spruce and balsam fir. The remaining trees provide enough shade to inhibit the growth of pioneer species such as poplar, birch and pine, and an abundant seed source is available from the adjacent seed trees.

Because the seeds will grow into trees with similar characteristics to that of the parent, it is important to select seed trees with desired characteristics such as straight trunk, rapid growth, etc. The practice of high grading (the logging of all the best individuals in a forest) is discouraged because the successive forest generations will continually inherit the characteristics of the poorer trees.

The disadvantage of natural regeneration is that the forest will be inclined to follow a succession found in nature, which tends to be less productive than a managed forest. Natural regeneration is often more expensive and time-consuming in terms of additional silviculture treatments. Natural regeneration, as many professional foresters will attest, does not always produce the desired species and when it does, the juvenile trees are usually improperly spaced and require additional planting or thinning.

Pseudo-Natural Regeneration

Pseudo-natural regeneration is the practice of assisting natural regeneration with the preparation of proposed regeneration sites. Site preparation includes the following: disposing of slash material, using mechanical disturbances to expose mineral soil, animal or chemical control of brush, and enhancing the quality, quantity and species of seed available. The best examples of pseudo-natural regeneration can be observed in clearings made for hydro and pipeline rights-of-way and on roadsides.

Often slash must be removed or burned, especially when falling trees on a lawn or cultivated field. Burning slash on organic soils or during the summer months is not recommended because of the risk of the fire escaping. Always consult with municipal or forestry personnel before burning to obtain information on fire conditions, need for burning permits, etc.

Natural conifer regeneration is evident along many roadways.

This modified front-end loader can be used to deck logs or to move slash off the field.

An effective technique for promoting pine regeneration in large, relatively flat clearcuts is achieved by dragging heavy chains or similar heavy implements behind a tractor to align and crush slash, expose mineral soil and bring the cones closer to the ground. Where slash accumulations are light and stumps are small, agricultural implements such as disks and ploughs may be used to expose mineral soil.

In very small woodlots, a rototiller can expose mineral soil and may be very satisfactory for stimulating natural regeneration. These mini-natural nurseries can produce lots of seedlings every year which can be used in a planting program.

Competition from grasses, weeds and other non-woody vegetation can be very detrimental to tree establishment. This vegetation may be controlled in several ways; herbicide application is the most common. Although herbicides can control the herbaceous competition, the application can be expensive, time-consuming and dangerous to the applicator and the environment. Herbicides should only be applied by certified technicians who have a complete understanding of the herbicide and its effect.

The grazing of sheep and other small-framed livestock is becoming recognized as a very important management tool in the control of herbaceous competition in conifer regeneration (SEE CHAPTER 8). The sheep essentially consume the surrounding vegetation, allowing early conifer release. Unfortunately sheep browse on the leaves and buds of poplar and other softwood trees and are not suitable for vegetation control for deciduous regeneration.

When no seed source is available for a particular regeneration site, as often occurs when farmland is returned to forest, alternative methods of bringing seed to the area must be used. To compensate for the incredible odds against a mature tree growing from a single seed, mother nature produces enormous quantities of seeds. A small backyard woodlot produces enough seed to reforest many acres of land.

The seed does not necessarily have to be in the raw state to provide good restocking results. A good way of providing seed to such an area is to take the delimbed branches with cones from a nearby logging operation and spread them on the ground. The branches hold the cones in place and the seed drops to the ground. Cones detached from the branches can also be spread, but strong winds may "drift" the cones into a large pile.

Raw seed may be spread, but be careful not to overseed because coniferous and deciduous seeds are very small. When direct seeding, it is often a good idea to harrow or lightly disk the area to bury the seed. A shallow seed burial hides the seeds from rodents and helps prevent the embryonic plant from drying out.

Seeds for conifer and deciduous trees can be procured in a number of ways, but the best is to gather natural seed. It is very important to collect seed from a local source because of genetic advantages. Trees adapt themselves to grow in particular soil and climatic conditions. The closer the seed source to the planting location the more successful the regeneration program.

Poplar seeds are easily gathered while in the cotton stage, just before dispersal in early summer. Other trees such birch and alder produce seed clusters that can be harvested in early autumn.

For conifers it is important to determine the seed maturity before cone collection begins by slicing the seed of several cones in half. If the seed is firm and the embryo has turned yellowish, the cones are ready for picking. Cone harvest usually begins in late August and continues into October before the cones opening on the tree and releasing seed.

Cones may be picked directly off a tree by using a ladder or platform attached to a front-end loader. This technique of cone collection tends to be more dangerous and labour intensive than other methods, but is the best way to ensure uniformity of progeny and obtaining seed from the tree with the most desirable characteristics.

The trees along this woodlot margin were rototilled, resulting in a large number of juvenile white spruce seedlings. The excess "wildings" are then dug up to restock other areas.

Cones may also be collected from the slash remaining from delimbing after a harvesting operation. The cone-bearing branches can be scattered over the proposed regeneration area or cones can be removed for seed production. Because the tree the cones are taken from is no longer standing, it is impossible to determine whether the cones come from desirable trees or not.

A shortcut to cone collection is by letting the squirrels do the cone collection for you. The squirrel's natural behaviour is to go from tree to tree dislodging cones and letting them fall to the forest floor for a retrieval and storage later. These cones can then be picked off the ground into ice cream pails or other receptacles before the squirrels can gather them up.

Because squirrels do not care where a dislodged cone lands, the cones from several trees will be mixed when they are thrown to the forest floor and genetic purity cannot be guaranteed.

Squirrels store cones for winter use by caching them in hollow logs and underground burrows. If you place a wooden box with a removable top and a small entrance hole to simulate a natural cone cache, squirrels will unwittingly gather cones from the forest floor and fill the box. The cones are easily removed by opening the box top and emptying them into a burlap sack or other container.

These cones are drying in the sun on a sheet of black plastic in the backyard.

Use discretion when harvesting cones. A single cone may contain a hundred or more seeds anda sackful of cones will usually provide many more seeds than necessary for a woodlot regeneration program. Always remember this "free" supply of cones is really part of the squirrel's winter food reserve. Removing all of the cones may result in starvation for the squirrel.

Once the cones have been collected, they can sold "as is" directly to private or government nurseries. These nurseries use sophisticated machinery for cone seed removal and often use the seed for inhouse seedling production. People who have the inclination can remove the seed from the cone at home.

Freshly picked closed cones will have a moisture content of up to 30% so store them for a short term in an open cardboard box or perforated container such as a burlap sack to allow air circulation through to prevent mould and mildew from forming. For longer term storage or seed removal, completely dry the cones to moisture contents of less than 12%.

Cones can be dried for storage either passively or actively. In passive drying, cones are spread on a black background and left to sun dry for a couple of days to a week. Active drying involves the application of an external heat source such as suspending cones placed in a burlap sack over a heating duct.

With a few precautions, cones can be dried very quickly in a conventional clothes dryer. Place cones *securely* in a flannel sack and then in a clothes dryer at *low* heat. Fill the dryer reasonably full with cone sacks to prevent excessive bouncing, which may rupture the sacks. The dryer action dries the cones, and the vibrations caused by the tumbling action remove the seed from the cone. Pine cones, because of the resinous coating that secures the seed inside the cone, often require slightly more heat to remove the seed from the cone.

The disadvantage of tumble drying the cones in the household dryer is that often the chaff produced in the drying process will appear in many loads of clothes afterwards. If your spouse is concerned, you can remove the seeds from passively dried cones by placing the cones in a clean ice cream pail or small garbage pail and shaking it vigorously.

Once the seeds are removed, they still have their wings attached. The wings aid in seed dispersal in nature but tend to be a nuisanceto the woodlot manager. Wings can be broken off by placing the seed between two pieces of cloth and rolling the seeds on a hard surface. The broken wings are much lighter than the seed and can be removed by alternatingly sifting and blowing on the seeds.

Balsam fir seed with wing.

Once the seeds are dried and the wings removed, they can be either planted immediately or stored for future use in a glass or other rodent-proof container.

Artificial Regeneration

Planting seedlings is usually the regeneration method of choice for many woodlot managers. Planting offers a number of advantages over other types of restocking programs. Planting will give a one- to five-year head start over natural regeneration and allows more control over species and tree quality. Last but not least planting allows you to space seedings at optimum distances to maximize growth and reduce secondary forest treatments such as thinning.

Seedlings used for planting can be obtained in a number of ways. The seedlings can be taken from the wild, they can be grown at home, or they can be purchased from a commercial nursery.

Seedlings taken from the wild can benefit the woodlot manager in a number of ways. Natural seedlings (wildlings) tend to be much hardier than the "pampered" greenhouse-raised seedlings which have been grown in conditions of optimum light, moisture and nutrients.

Wildlings are locally grown and are usually genetically better suited to grow in that particular microclimate. Often wildings are a byproduct of juvenile thinning and can be obtained for free. If wildlings are to be removed from public land or road allowances, a permit for their removal must be obtained first.

Wildling production is an excellent conservation project for urban woodlot owners. Wildling production can be stimulated by exposing mineral soil around the base of a cone-producing tree with a rake or a rototiller. (Be careful not to damage the roots of the seed trees!) Seeds may be spread into the newly disturbed earth, or let nature provide the seed. Often this type of seedling production produces amazing results where many plants must be removed every year to prevent overcrowding.

Wildlings should only be transplanted in the early spring or the late autumn. The trees are in a state of dormancy at this time and are less prone to desiccation caused by damaged roots unable to supply enough water for evapotranspiration.

When digging wildlings by hand, select trees less than .6 m (2 feet) tall. Shovel cuts should be no closer to the stem than the tip of the farthest branches, otherwise excessive root damage will occur. Larger trees can be dug with mechanized equipment such as a tree spade, but the economics of such a venture make this type of operation only feasible for ornamental planting.

The correct procedure for digging and transplanting trees is as follows: 1) with a shovel, cut a square around the tree as big as the largest outside branches, 2) with the shovel as a lever, raise the tree from the ground—do not pull on the stem to assist in raising the tree, 3) wrap the base of the tree in wet burlap or cardboard, 4) place the tree and burlap into a sheltered container or truck box—driving long distances with the seedlings exposed to the wind will result in severe desiccation and high mortality rates, and 5) plant immediately.

Seedlings may also be purchased from commercial nurseries and greenhouses. The advantages of commercially grown seedlings are threefold. First, for sizable reforestation projects, it is possible to purchase large numbers of seedlings at minimal cost without the inconven-

A window sill with a grow light is an ideal place to grow containerized seedlings.

ience of finding and transporting natural seedlings or growing your own. Second, it is easy to introduce non-native species. Third, commercially grown species usually have been selected from genetically superior individuals and may exhibit superior growth characteristics than local trees.

The seedlings are grown in plastic or styrofoam trays filled with high nutrient soil called "plugs." The plugs are only suitable for growing seedlings up to one year of age and are the seedlings of choice for large commercial reforestation programs. The rich plug soil can give the seedling a survival advantage when attempting to reforest nutrient-poor sites.

An alternative to buying plug stock is to grow your own. Plastic or styrofoam seedling trays are available from wholesale greenhouse supply companies. The trays are inexpensive and with proper care will last for years.

Fill the seedling trays with soil. Potting soil is usually the best for seedlings. You can purchase the potting soil from your local hardware store or you can make it home by preparing a homogenous mixture of 1/3 garden soil, 1/3 peat moss and 1/3 vermiculite. Once the planting trays are filled and compacted with the potting soil, the seeds can be planted.

Before planting the seeds, it is necessary to simulate the natural seed cycle of fall planting, winter dormancy and spring germination. The simulation procedure for spruce seedlings is as follows: 1) soak the seeks overnight in a pail of water, 2) remove the seeds and place them in a plastic bag or other sealed container and place them in the refrigerator for three to four weeks, 3) remove the seeds from the plastic container and place three seeds on top of the potting soil and cover with coarse sand to a depth of .5 cm (1/4 inch), and 4) water lightly and cover the trays with clear plastic wrap to keep in the moisture and place the trays in a warm well-lit room.

Small homemade transplanting sleds can transplant hundreds of container seedlings per hour.

Once the seedlings have germinated and are poking through the sand, remove the plastic wrap and water the plants every couple of days or so. If more than one seed germinates, leave the tallest, most vigorous growing stem and cut the others with a pair of scissors. Make sure the seedlings receive at least 16 hours of light daily. Provide additional light by hanging fluorescent tubes over the tree seedlings. Home-grown plug stock can be seeded in the same manner as commercially grown plug stock.

Plug stock is ideal for machine planting of agricultural land and irrigated plantations. Plugs may also planted in natural areas with an elongated track shovel or a potipuki.

A tractor and a planting sled can plant several thousand trees per hour at a uniform spacing, which is necessary when drip irrigation is used. The plugs are planted at 30 cm (1 foot) spacings to match the emitter spacing of the drip irrigation hose. These trees are grown under irrigation for two years (until the branches of adjacent plants touch) and are then transplanted as bareroot stock.

With a potipuki it is possible to plant plug seedlings without bending over. The tool is pushed into the ground and the tree is dropped in at the top of the tube (left). The retractable jaws are opened and the tree remains in the hole (middle).

At this provincial tree nursery (below), bareroot white spruce seedlings are sorted and bundled for farm shelterbelts. A slight depression around the newly planted trees will hold water and increase rate of survival (right).

Sites with a profusion of competing vegetation are often better off when seeded to larger bareroot stock. The bareroot stock is usually three years or older and is so called because the dirt is washed off the roots immediately after the seedling is dug. The removal of the dirt makes transportation of bareroot seedlings cheaper and easier.

Bareroot seedlings, however, are extremely susceptible to desiccation and must be planted immediately after the shipment is received. Bareroot seedlings are also well suited to planting with mechanical planters and are often used to establish farm shelterbelts.

This shade tolerant spruce seedling was planted on the north side of this beaver-logged aspen.

Site preparation for newly planted trees varies with the type of tree, the method of planting and the land contours. Most shelterbelts planted with a mechanical planter on farmland require well-cultivated soil to allow for easy planting. Steeply sloping sidehills should not be cultivated to prevent water erosion. Grassy sidehills should be mowed to facilitate easy straight planting and sprayed with a non-selective herbicide applied to prevent competition with other vegetation.

Planting with a spade is suitable when small numbers of trees or uneven terrain make mechanical planting unfeasible. The proper steps for planting are as follows: 1) dig a hole or make a cut with the spade to a depth of about 15 cm (6 inches), 2) place the tree in the hole to the depth where all the roots will be covered by dirt, 3) remove the spade and tamp the dirt around the roots creating a slight depression to collect water, 4) give the tree 2 to 3 litres (.5 gallon) of water or until the depression becomes full, and 5) water the trees once a week for the first three weeks until the roots develop.

When replanting a harvested area it is important to give the newly planted tree the best opportunity for survival. Planting the seedling close to the stump of a harvested tree will increase the survival rate. The decay of the stump and the roots will provide nutrients to the seedling and root decay will leave a tunnel through the soil for water uptake. Pioneer species should be planted on the south-facing side of the stump, while shade tolerant trees should be planted on the north.

Newly planted trees, especially those planted in open places such as farm shelterbelts, will have substantially higher growth rates when protected from drying forces of the sun and the wind. A snow fence or other type of protection will prevent excessive water vapour exchange from the leaves and prevent the drying out of the soil.

This modified planter (left) rototills, plants and waters all in one operation. This hillside (right) is susceptible to water erosion and therefore cultivation prior to planting is not suitable. Instead a path is mowed and sprayed with a non-selective herbicide prior to planting. The measuring stick and rope ensure proper row spacing.

The six-year-old lodgepole pine seedling (far left) protected on the south by a cattle wind fence exhibits double the growth as a similar aged lodgepole pine seedling (left) planted in an exposed area.

Chapter seven

MARKETING FOREST PRODUCTS

Traditional marketing of forest products was once restricted to selling standing or decked timber. Although this practice probably still represents the largest revenue-generating mechanism in the woodlot, many managers are finding economic opportunities in marketing specialty forest products aimed at small niche markets.

One advantage of the longevity of a tree rotation over other types of crops is that harvest need not commence in any given year. Market cycles are largely driven by supply and demand. When supply is high and demand is low, prices are also low; the reverse happens in the opposite situation. If personal finances permit, a woodlot owner can reduce or eliminate the volume of lumber harvested during a period of low prices.

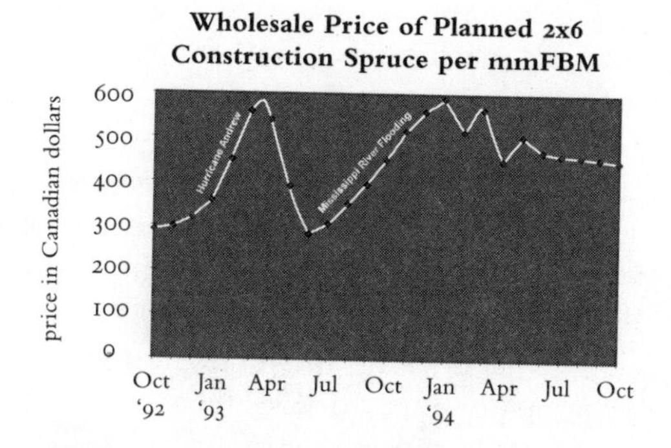

Lumber peaks are caused by natural disasters in the United States.

House-building in the United States is the single largest industrial sector responsible for creating lumber demand in Canada. Housing starts generally increase in an expanding economy, and tend to decrease during recessions. The demand for lumber and, hence, the anticipated price, can also be expected to be strongest in an expanding economy. Short-term house-damaging catastrophes such as hurricanes and floods can also be a positive influence in creating lumber demand.

Pulp log prices are also affected by shifts in the economy, and also tend to be higher in an expanding economy. Pulp prices are not affected by short-term catastrophes but can be influenced by short-term increases in paper use such as during a national election.

More recently, prices for private wood have been increasing, especially in B.C. Many mills have had their annual allowable cut from public lands decreased. The mills, desperate to keep production at full capacity, have paid enormous premiums to private landowners as far away as Saskatchewan.

Large mills usually purchase wood from private land in one of three ways: as standing, decked or delivered timber. Which purchase agreement is best for a particular woodlot depends on volume of timber, equipment and labour on hand, species and type of harvest.

Selling standing timber is the easiest management option in time, labour and capital required. Usually the buyer is responsible for all of the harvesting operation including falling, delimbing and road construction. Standing timber contracts (STCs) may also include clauses covering scarification and/or reforestation.

The standing timber contract, while minimizing inputs by the woodlot manager, also minimizes revenue generated because someone else must be paid for the harvest. STCs usually involve large clearcuts where it is profitable to use large machinery. It is still your responsibility to clearly define cutting specifications and obligations to the logging contractor (SEE CHAPTER 4).

The STC should outline cutting deadlines, how the wood is to be measured and the price per unit of timber received. Timber sale boundaries should be clearly defined to prevent trespassing onto neighbouring land. In the case of a selection cut, remember to specify dbh and species of trees to be harvested.

It may also be advisable to retain the services of a lawyer or forester with timber contract experience to negotiate type and amount of compensation for unauthorized damages or unfulfilled contractual obligations. It is critical that the logging contractor has a Workers' Compensation Board (WCB) policy that covers all personnel involved in the harvest operation.

These white spruce have been decked and are awaiting transport to a processing facility.

In urban woodlots, often only single trees are removed when they die or present a hazard to people or property. The safest method of harvest is to call in tree removal specialists. Again, make sure the company has proper insurance and detail how the tree(s) is to be felled and how it is to be cut up once it is down.

Decked timber refers to trees that have been fallen, delimbed and piled in a landing, field or other truck-accessible area. The timber is then processed on site or trucked to processing facility such as a sawmill or pulp mill.

Selling decked timber offers a number of advantages. It allows the woodlot manager complete control over the harvest operation including road building, landing construction, falling, delimbing and decking. It is usually the most cost-effective alternative in farm woodlots where excess capacity of labour and machinery may be used. Often the most important consideration is the investment of additional time, labour and capital, which is usually justified by a higher selling price.

The woodlot manager usually is able to adjust the woodlot harvest schedule to avoid conflicts with other non-forestry operations such as farming. By harvesting during the winter, farm woodlot owners can employ machinery and labour that would normally be idle. By adjusting the timing of tree harvest and sales, you also have the advantage of maintaining a steadier cash flow.

Most farm woodlot operations selling decked timber deal with small volumes. The harvest operations do not involve specialized forestry equipment. The falling and delimbing operation is accomplished using chainsaws or tractor-mounted logging attachments. Logs are dragged to the landing behind a tractor and piled into decks using a front-end loader. The logs can then be scaled at the landing before processing or, as is most often the case, loaded on trucks and then weighed or scaled at the mill.

Obtain market and price guarantees for the timber *before* the harvest operation. Pre-selling the timber gives you the advantage of avoiding short-term price fluctuations and gives added insurance that the harvest operation will be profitable.

With delivered timber, unit price also includes delivery to the mill or other loading facility. Delivered timber contracts are usually only an option to woodlot managers who have large enough trucks for delivering tree-length logs, subcontract log delivery or deliver small quantities of cut logs such as firewood.

The entire harvest in small woodlots is often completed by one or two people. It is essential that each person receives proper training before the initiation of harvest, including chainsaw and first-aid certification. All personnel should also be covered by a Workers' Compensation Board (WCB) insurance policy as many mills will refuse to accept delivery unless proof of proper WCB coverage can be produced.

The practice of selling roundwood to a processing facility such as a sawmill or pulpmill is changing as woodlot managers realize the economic benefits to manufacturing their own value-added products. Sawing of lumber and firewood still represents the most common form of primary processing, but other value-added industries such as furniture making are finding their own niche markets.

Small-scale sawing operations have existed in this country almost from the time Europeans first set foot in Canada, so it comes as no surprise that many of today's woodlot owners continue this tradition. Although the first sawmills had machinery driven by water or animals, the modern sawyer can choose from two types of small portable mills: the circular mill or the bandsaw mill.

The circular mill has long been used by farmers as a good reliable method of sawing lumber into rough boards or planks. The circular mill cuts the lumber using a large round-toothed blade, much like a giant version of a skill saw. The blade can be driven by a sawmill-mounted engine, but usually the power supply comes from the power take-off from a farm tractor. Logs are mounted on a steel carriage and then propelled into the rotating blade. The lumber from these mills can be used for a

A typical circular saw mill in operation.

The board from the circular saw mill (below) needs to be planed (right) before it can be used in many applications.

wide variety of building applications where a planed surface is not necessary, such as framing, trusses, fencing, etc.

In 1982, bandsaws were introduced as an alternative to the circular saw. The bandsaw mill looks and operates much like the bandsaw in the butcher shop. A small, sawmill-mounted gas or electric engine powers the loop blade around two guide wheels. Logs are loaded on the sawing deck and the blade is propelled through the stationary log.

Each type of saw has advantages and disadvantages. The circular mill has a much higher productivity, requires less blade maintenance and is able to saw extremely knotty timber. The circular saw, however, is longer, heavier and more cumbersome to move and set up and requires more horsepower to drive the blades. The cut lumber is coarse and must either go through a secondary planing process or be used where smooth lumber is not necessary.

The bandsaw, because of its smaller size and low required power input, tends to be safer and quieter to operate. The thinner blade thickness (kerf) results in less sawdust and up to 25% more lumber from a single log. The bandsaw can usually saw larger logs and the finished lumber has a much smoother finish and may be used in a wide variety of applications without planing.

Although portable mills cannot compete profitably with large sawmills for standard-sized spruce and pine lumber, small mills can carve their own niche custom cutting unconventional sizes, species and products.

Portable mills, especially bandsaw mills, can cut odd lengths of lumber that would normally be discarded by conventional mills. This allows the sawing of beams up to 9 m (30 feet) in length for home and barn construction. Logs may also be slabbed on two or four sides for wall material in log buildings. Small lengths of wood may also be sawn into shingles.

Bandsaw mills, because of their small size and quiet operation, have applications in populated areas. Many urban dwellers grow valuable species for their ornamental value only. Once these trees die or become a hazard, they are felled, bucked and either hauled to a landfill or burned as firewood. Often the tree owner will give the tree away to save the cost of hiring someone to remove it. Many ornamental trees have unusual and desirable qualities and command premium prices from specialty lumber buyers. Bandsaw mills also can make tapered cuts that allow "trash" timber to have commercial applications such as sidings and roof shingles.

Once the sawing is completed, it is important to stack and store the lumber properly. If the lumber is still "damp" (moisture greater than 15%) then it must be "dry piled," otherwise mould and mildew will begin to deteriorate it.

Dry piling is a method of stacking lumber to allow the sun and wind to remove moisture from the lumber. The lumber dries evenly without excessive warping or cracking. The disadvantage to dry piling is the length of time required before the lumber is usable. The complete drying process depends, of course, on initial moisture content of the wood, weather condition, etc., but usually takes from two to four months.

Bandsaw mills are quiet enough to be suitable for sawing operations in urban areas (left). Lumber can be cut from logs with a chainsaw equipped with special attachments (right).

These boards (left) have been improperly piled and are uncovered, and their quality has begun to decline owing to mould and mildew formation (right).

To dry pile lumber, 1) lay treated lumber, steel or concrete rails on the ground (in a building or shed if possible) to elevate the first layer of lumber, 2) lay the damp lumber side by side on top of, and across, the ground rails, 3) place spacer boards 2.5 cm (1 inch) thick and 10 cm (4 inches) wide at 30 cm (12 inch) intervals across the first layer of damp lumber (damp lumber may be used as spacer boards), and 4) repeat this process until the lumber is completely stacked and cover with a waterproof tarp if a roof is not already over the lumber.

Lumber can be dried more quickly by using a lumber drying kiln. Drying kilns are usually of two types, forced heat or solar. Forced heat kilns are commonly used by the larger mills and use heat by burning either natural gas or wood byproducts. Large fans push the heated air past the stacked lumber, which dries very quickly, usually in a couple of days. The disadvantage of forced air kilns is the high energy requirements to heat the air.

A solar kiln is more than sufficient to dry all the lumber from most woodlots. A solar kiln looks and operates much like a greenhouse and in fact can be used as one in the off-season. For maximum drying

These boards are properly dry piled. Each layer of damp wood has spacers at regular intervals to allow the prevailing wind to evenly dry the lumber (left). The entire pile is safely stored away from the rain and snow under a pole shed (right).

The inside (left) of a homemade solar wood kiln. The kiln faces south and is protected from pevailing winds by a large building. A woodworker (right) with the initial wooden products.

potential, it is best to face the kiln south preferably protected from the prevailing winds by a barn, shop or other building. Hot, moisture-laden air from the kiln interior can be vented outside by using a small electric fan. Drying times for lumber in a solar kiln are usually three or four weeks.

The apex of secondary processing is the manufacture of finished products in a woodworking shop. Traditional products such as tables, chairs and lamps, and some non-traditional products such as canoes and candle holders, can easily be built in even the most rudimentary woodworking shop. The secondary processing may just be a hobby that supplements the family income or it may become a rewarding full-time occupation.

Small diameter logs, which are not economically feasible to saw, can easily be processed into fenceposts or rails. Some tree species, such as cedar and tamarack, already contain their own natural preservative and can be used as posts and rails right out of the woodlot. Most other species, however, require a chemical treatment to enhance resilience to decay.

There are essentially three types of chemicals available to the woodlot owner for fencepost preservation: 1) poisonous materials dissolved in solvents (pentachlorophenol), 2) chemicals dissolved in water (copper sulphate, zinc sulphate), and 3) toxic oils (creosote, tar). By far the safest and easiest method of treating posts is the displacement of sap with copper sulphate.

The sap displacement process uses freshly cut and peeled stems immersed butt down in a solution of 1 kg (2.2 pounds) of copper sulphate mixed in 6 litres (1.3 gallons) of water. Leave the posts in the

This tamarack fencepost was treated with copper sulphate as a preservative.

container until the blue colour rises to the top of the posts, usually within 6 to 48 hours. The average 2 m (6.6 foot) post with a 10 cm (4 inch) butt diameter will use approximately 2 litres (.5 gallon) of solution.

Although copper sulphate is inexpensive and readily bought over the counter at many hardware/farm supply stores, it also has its dangers. Copper sulphate is corrosive and proper safety equipment such as rubber gloves and eye protection should be worn at all times when handling this material. It is also important to use a plastic or fibreglass immersion tank, as steel tanks will corrode very quickly and develop holes within a day or two. For small quantities of posts, a large plastic garbage can will suffice.

The life expectancies of untreated and treated posts are shown in Table 3.

Salvaging waste wood products can be done by variety of methods. The most common method is to cut a defective board into a smaller board of premium quality. Quality boards may also obtained by recognizing valuable wood in buildings slated for demolition or wood products intended for the dumpster. Last, but certainly not least, many waste wood products have valuable uses.

Canadian lumber grading rules stipulate the type and amount of defects a board may have before it is reduced in grade. Often mills will cut only a particular size or sizes of board on a given run. Boards will be reduced a grade or two even if only a very small portion of the board is

Kind of wood	***Untreated life***	***Treated with copper sulphate***
Cedar	20–30 yrs	30–40 yrs
Tamarack	10–15	20–25
Pine	8–12	15–20
Birch	5–10	12–20
Willow	5–8	12–15
Spruce	4–6	10–15
Poplar	3–5	10–15

Table 3: Life Expectancies of Untreated and Treated Posts

defective. Because large mills make their profit on large volumes of wood, it is not economical to separate individual boards and defective boards are placed in a stack (lift) and sold at a reduced price.

Lifts of degraded boards may be purchased at a very low price and have the defective portions of the board sawn off and resold as "number 1" at a premium price. This type of operation, while potentially very profitable, is also very labour intensive because of the necessity to continually handle individual boards.

Another type of resaw operation involves sawing square edges (edging) on slabs or slab boards. Slabs have one sawn side, and slab boards have two sawn sides. Slabs and slab boards come from cuts on the outside perimeter of the log.

Large milling operations grind slabs into chips that are then used on site as a fuel source or sold for further processing to either pulp or oriented strand board mills. Slab boards are fed through an "edger," which cuts the bark portions off, leaving a narrower board with four sawn edges. Many small mills do not have chipping or edging facilities and slabs and slab boards can be obtained at a very low price.

For very small operations, an homemade edger can be built out of a radial arm saw. A track and carriage can be made using straight pieces of sawn lumber. Slabs and slab boards are attached to the carriage using clamps and are fed through the saw. Because only one edge of the slab board can be sawn at one time the productivity of this type of edging operation is usually restricted to fewer than 60 boards an hour. This type of operation can be profitable if slabs and slab boards can be obtained at a reasonable price.

This fence was built of slabs that have been edged. The original slabs were to have been burned.

After edging, slab boards are bark-free and can be sold as regular lumber. Slabs are used "as is" as wind breaks around animal pens or tree plantations. Edged slabs may also be used with the bark remaining or removed as either fencing or siding on garages, tool sheds, or in "cowboy" saloons, to give a rustic look.

Salvaging lumber from old houses and barns can be a very popular and competitive business, particularly if valuable types of wood have been used. Starting such a business, however, requires prior knowledge of building demolition, evaluation of wood soundness and lumber pricing and should not be attempted by inexperienced people.

Many opportunities exist in salvaging wooden boxes, crates and pallets, especially those imported from Asia. These discarded containers may either be used again or resawn into different products. Containers or pallets imported from Asian countries are often made, because of the lack of an alternative, out of mahogany or teak, which can command very high prices as a specialty wood.

Larger mills are now required by law in many provinces to ship wood shavings and chips to secondary processing or pulping facilities. Portable mills can use these shavings as livestock bedding, as mulch between trees in commercial orchards or as a fuel for supplemental home or shop heating. Aspen chips are particularly good as a smoke source when smoking meats and fish.

Sawdust can also be marketed in number of ways to take advantage of its absorbent properties. The sawdust can be used for animal and pet bedding, sold to service stations or farm shops to soak up oil and gas spills and occasionally is used in the oil-well drilling industry as a filler when downhole circulation is lost. Sawdust is also an excellent filler to use in potting soil and gives garden soil gives additional tilth and water retention capabilities. When mixing sawdust with soil, it is important to add extra nitrogen fertilizers to help the soil organisms decay the wood fibre.

A radial arm saw was used to edge these slabs for the fence shown opposite.

These large saskatoons taste great in jams or jellies, or right off the tree.

As well as providing fibre for lumber, pulp and firewood, the woodlot can provide a cornucopia of marketable products. Markets exist today for a variety of unusual forest products including herbs, botanicals, berries, mushrooms, syrups, flowers, branches, bark and oil extracts.

The rising cost of medical care, especially in the United States, has resulted in an annual growth rate of over 10% in the consumer use of herbal medicines. In North America alone this amounts to over one billion dollars of sales annually. Plants such as bearberry, yarrow and fireweed are just a few of the dozens of forest plants that have an existing market.

Berries such as saskatoons, chokecherries and wild rose hips are gaining popularity beyond their use in jams and jellies. These berries, derived from natural sources or from plantations in agroforestry operations, are finding their way into specialty beverages, wines and specialty teas.

Branches and cones can be made into Christmas wreaths (left) or they can be combined with dried flowers to make "no-occasion" decorations. This type of wood sculpture (above) is called "intarsia" and commands a very high price in specialty craft shops.

The forest is a wonderful place for quiet contemplation and serene meditation (left). Camping is fun even when the comfort and safety of the house is only minutes away (right). These pictures show better than any words some of the many joys of owning and maintaining a woodlot.

Branches and cones have traditionally been made into wreaths and other ornaments for the Christmas season but the floral industry is showing significant interest in year-round "drieds" or "permanents." Interest is particularly high for products with the natural "wildcrafted" look. Floral wholesalers are developing full product lines that depict a rustic "hunting lodge" look, which also includes a number of popular products such as twig and branch furniture, stump candle holders and bark containers.

Many types of art forms use wood as the primary ingredient. Intarsia involves making a picture by cutting out parts of an original design from different colours of wood, polishing them and combining them in an intricate inlay that fits together like pieces of a puzzle. Other types of more familiar wood artwork include duck carvings and jewellery boxes.

Oils extracted from wood and needle biomass are being tested in the world's major perfume and fragrance industries. Studies have indicated that employee productivity is increased when pleasant-smelling "forest" aromas are added to the workplace atmosphere. White spruce, black spruce, jack pine and balsam fir offer the best opportunities for this type of product.

The woodlot can also be a great storehouse of value without harvesting anything. Managed and unmanaged woodlots can add considerable value to residential or rural property. Trees beautify the home, increase crop production and offer spiritual refuge from hectic day-to-day living in the modern world.

The trend towards urbanization has created a demand for natural, outdoor recreation. The recent renewed interest in camping, "country vacations," and bed-and-breakfast facilities is testimony to the desire for all types of people to "get back to the country."

The woodlot manager no longer needs to think of the woodlot as solely as a miniature fibre factory. Traditional forest products may be harvested and marketed in complete harmony with other recreational and economic aspirations. The woodlot is a great place to go walking, birdwatching, skiing, horseback riding, camping and picnicking. With proper management and an eye for the future, the woodlot provides a continual source of entertainment and financial benefits now and for generations to come.

Tree Trivia

Think of tree products and what do you think of? Lumber and paper. Listed below are some of the products trees provide.

PULPWOOD

Tall Oil
paints
laquers
soaps

Turpentine
paints
polishes
cleaning fluids

Lignin
soil conditioners
plastics
tanning materials
drilling mud additives

Cellulose

Chemical Products
rayon
cellophane
sausage casing
explosives
photo films
celluloid
shatterproof glass
sponges
imitation leather
artificial hair and bristles
moulded plastics
phonograph records

Fibre Products
newsprint
wrapping paper
paper bags
*books (*Woodlot Management*)*
writing paper
tissue and toilet paper
absorbent paper
wallboard
boxboard
hardboard
insulation board

SAWLOGS

Shop and Factory
boxes
cabinets
caskets
flooring
furniture
millwork
musical instruments
boats
sporting goods

Construction
mining timbers
joists, beams, columns
planks
sills
laminated timber
boards
dimensional lumber
finished lumber
railway ties

Mill Waste
slabs
fuel, lath, pulp
particle board
edgings
mouldings, dowels
mop handles
sawdust/shavings
animal bedding
insulation

VENEER LOGS

plywood, matches, toothpicks, doors

BOLTS

barrel staves, handles, shingles, pallets, charcoal, distillation products (tar, wood creosote, wood alcohol, acetic acid)

SAP

balsam-glass, cement, varnishes, spruce chewing-gum, drugs, confections

Chapter Eight

ANIMALS IN WOODLOT MANAGEMENT

Beaver Logging

In today's world, where it seems there is a perpetual struggle between industry and the environment, beaver logging is a perfect example of how humans and wildlife can co-operate for mutual benefit. The well-managed woodlot provides a colony of beavers with a continuous supply of food and building material. In return, the beavers, as a result of dam construction, create ponds that provide water for wildlife, forest protection and recreation, and provide substantial quantities of logs for sawmills or pulpmills.

Beavers are the largest rodents in North America and the second largest in the world behind the capybara of South America. Mature beavers can grow as large as 45 kg (100 lbs) but the average adult usually weighs in at around 25 kg (55 lbs). Skeletal remains indicate that ancient beavers grew to sizes in excess of 200 kg (450 lbs).

Water is the factor most critical to beaver survival. Ideal beaver habitats are small lakes, ponds and meandering streams in forested areas. Often young adults expelled from an existing colony travel several kilometres overland in search of suitable habitat.

Beavers live in colonies that consist of the mother and father and two generations of young. Beaver babies, called "kits," are born in early May in litters that usually average three to four. The young beavers are ousted from the colony after the second year and then must fend for themselves.

The beavers' preferred diet comprises primarily the leaves, twigs and bark stripped from the branches of fallen aspen trees, although just about any tree will do. After the branch bark has been eaten, the remaining sticks are used in dam or lodge construction. The larger pieces

This staged shot illustrates the beaver dam, an engineering marvel that has gained the beaver international recognition.

of trunk are left to rot. The average beaver eats about 1 kg (2.2 pounds) of food a day, and a colony of beaver may knock down as many 100 mature trees in a single year.

The beavers' dam-building instinct is probably its most distinctive feature. The dams are built primarily from barkless sticks and small logs woven together with mud and other debris. Most dams are under 50 m (160 feet) long and the resulting beaver pond comprises less than 1 to 2 hectares (5 acres), although some dams are over 600 m (2000 feet) long with ponds of more than 50 hectares (125 acres).

Beaver ponds benefit wildlife and the woodlot manager. The ponds formed behind beaver dams provide drinking water for forest animals, nesting and feeding sites for ducks, and habitat for fish. Beaver ponds provide the woodlot manager with adequate water supply for fire protection, a swimming hole in the summer, a skating rink in the winter and a year-round location to enjoy nature.

Beavers also provide economic benefits to the landowner. Beaver logging is a salvage operation made in heaven: the woodlot manager removes the only part of the tree that the beavers cannot use, which coincidently is the only part of the tree the manager *can* use. Beavers cut down, delimb and often debark the tree, leaving the landowner with the relatively simple task of hauling the log out of the bush.

Tree removal is easiest when farmland or roads are next to the beaver-logged area. In this case it is possible to travel along the entire length of the stream or creek pulling beaver-felled trees up a short embankment. Often the trees have fallen directly on the road or field where the logs can be loaded directly on the logging truck.

The beaver has taken all it can off of this tree, and the leftovers are ready to haul to the sawmill or pulpmill.

Beaver-logged areas also require little cleanup effort. The leaves, twigs and branches that ordinarily would have been burnt or left as slash in a traditional logging operation are either eaten by the beavers or used to build dams and lodges.

Beavers are also important in forest succession. In a mixed-wood forest, beavers will selectively cut pioneer species such as poplar, leaving conifers and other deciduous trees intact. Cutting poplars opens the forest canopy, thus allowing the release of the conifer understorey.

Salvaging timber from beaver-logged areas can present some unique problems. One must be extremely careful, for example, when skidding trees from a beaver-logged areas. The pointy stumps resulting from beaver cutting can damage tires and cause serious injury to horses and humans. As an added precaution, skid trails should be well marked with fluorescent flagging tape and potentially dangerous stumps cut to ground level.

Another danger to be aware of when salvaging timber from beaver-logged areas is the large number of hung-up trees. Contrary to popular belief, beavers are unable to control the fall of a cut tree, hence a large number of trees lodge on other trees. The proper method of removing lodged trees is discussed in Chapter 4.

This beaver-logged area contains virtually no slash or other woody debris and is ready for reforestation.

Occasionally, beavers will abandon a partially cut tree. Incompletely cut trees are extremely susceptible to blowdown and premature falling caused by other types of vibrations such as a movement of machinery. These trees pose a greater hazard the more the beaver has gnawed off. To prevent accidental falling, partially cut trees should have felling completed by mechanical means before entering the area for log removal.

Often when initially attempting to salvage beaver-logged areas, a woodlot manager is faced with an apparently insurmountable task of cutting a trail through a tangle of beaver-felled timber from years gone by. Although it may not appear to be worthwhile to waste time cutting through wood with limited or non-existent commercial value, this work will produce big dividends in the long run.

Because the felled timber discourages ground travel, trails provide wildlife, especially larger ungulates, access to browse that originally would have been avoided. Because most or all of the standing timber has been felled, it is also the best time to build trails that will most benefit your woodlot goals. Lastly, the downed timber represents a large fuel source for a ground fire, and therefore trails are important for forest protection.

If beavers are desired in a particular forested area that does not support a population, relocation of individuals from another area is possible. First ensure that a source of water is available. Almost any size of stream or pond can support a beaver population, or an artificial pond can be dug inexpensively and will be suitable for beavers and for forest protection and recreation.

To remove hazardous trees, complete the initial beaver cut with a chainsaw as described in Chapter 5. Gnawing the remainder of the tree is not recommended.

This deck of beaver-logged trees was harvested from the author's woodlot. The paper in this book was made from beaver-logged trees.

Finding a beaver to introduce to a forested parcel is likely to be the easiest job of all. Recent low fur prices have reduced the annual fur harvest, resulting in a beaver population explosion. In many places in Canada, beavers and their dams have created havoc by flooding roads, rail lines and fields. Often beavers are regarded as a nuisance in urban areas, cutting down trees in yards and parks. A request to the local fish and wildlife officers will likely yield as many beavers as desired.

Although there are many advantages to beaver in woodlots, there are also some drawbacks. Dam construction may cause flooding of bridges and areas containing valuable stands of trees or field crops. Small woodlots, especially those in urban areas, may be in danger of being completely levelled by beaver activity. Even larger woodlots may be in peril if the rate of beaver cutting far exceeds what can be replaced.

Obviously, a beaver will cut trees occurring in the riparian zone closest to the creek or river they inhabit. Although the area cut by beavers is usually too small to create significant damage alone, when combined with areas logged by humans, serious erosional damage can occur. Therefore, extreme care must be taken by human loggers to ensure that a buffer zone of trees is maintained between human- and beaver-logged areas.

Beaver overpopulation can cause water contamination resulting in onset of the disease *giardiasis*. This disease, also called "beaver fever," causes extensive diarrhea and abdominal cramps. People swimming in beaver ponds with this type of bacteria present may contract this disease. Outbreaks have occurred in several major centres, including Banff, Alberta, and Penticton, B.C.

Problem beavers can be removed in several ways. The simplest way is to destroy the animals by hunting or trapping. No hunting or other licence is required to destroy beavers on private property as they are referred to as nuisance animals in all provinces. Beavers removed by such methods need not be wasted as their fur has some value and the meat can be made into some tasty dishes.

Beavers may also be live trapped and transplanted, but this operation should be left to experienced personnel. Most municipalities have experts in beaver removal, but because of the expense and lack of suitable areas to transplant them, most beavers are destroyed.

Tree Trivia

Interesting beaver bowel facts:

- *Beaver manure is primarily composed of "saw dust"*
- *Without water to "poop" in, a beaver will become constipated*
- *Beaver manure disintegrates rapidly in water, so beaver manure is not usually a good sign to identify the presence of beaver.*

Sheep Vegetation Control

Sheep controlling competitive vegetation in the woodlot is another way animals work with humans for mutual benefit. The newly planted woodlot provides the sheep with forage while the sheep provide an environmentally friendly method of removing competitive vegetation allowing for early release of conifer seedlings. As well as saving money for the woodlot management program, the sheep produce valuable byproducts such as meat, wool and manure.

As well as grass, legumes and weeds, sheep also browse on the leaves of small shrubs and willows.

Sheep, however, cannot be used for vegetation control on all reforestation programs. The sheep's diet comprises about 80% forage (grasses, legumes and weeds) and 20% browse (leaves, buds and stems of deciduous trees). Their diet makes them beneficial on replanted conifer forests, as they remove all competition including new deciduous growth, but on a newly planted deciduous forest the sheep will consume both volunteer and planted saplings.

The sheep grazing this reforested clearcut are in top condition. The Suffolk are the black-faced sheep.

The removal of the forest canopy provides ideal growing conditions for many types of weeds, grasses and brush. The rapid growth of the vegetation complex make some breeds of sheep unsuitable for grazing on reforestation programs. Breeds such as the Ramboulet, which have long fine wool or wool on the face, tend to collect twigs and other debris. Smaller breeds such as the Finnsheep may run into difficulty manoeuvring over the undulating slash cover terrain.

Suffolk is the best breed of sheep to use on woodlot grazing programs because of their tough hooves, heavy-boned legs and thick muscling. The Suffolk is less susceptible to eye problems because the short hair on the face and head can pick up twigs and other debris. This breed is also very adaptable to cold wet weather, is good at climbing through heavy bush and slash and is the most aggressive grazing sheep available. Suffolk are referred to as "heavy" sheep, meaning they do not frighten easily and will settle and graze in areas where other sheep would not.

Because conditions on a cut block tend to be more severe than on a normal pasture, only the animals that are in top shape should be considered for reforestation grazing. All sheep used in a grazing program should be inspected, vaccinated, pregnancy tested, dewormed, foot bathed and tagged before being turned out to graze.

Any sheep over two years of age can be used, but animals with a full set of teeth (five years old) are usually best for grazing reforestation programs. Juvenile and sheep under 35 kg (80 pounds) should not be used, nor should sheep that are sick, lame or do not appear to be in top condition.

Commercial grazing programs use mature dry open ewes for several reasons. The forest forage provides an excellent backgrounding summer ration and allows the animals to be in top condition for the autumn breeding season. Pregnant ewes tend to be more frail and mixing sexes on a grazing block is usually not advisable especially when using breeds such as Romanov, which are known for their off-season breeding capabilities.

Border collies are the herding dogs of choice on the cut block. They are hard working and gentle with sheep.

If the grazing area is near permanent stock-handling facilities and corrals, additional equipment is not necessary. Remote sites may require lightweight portable aluminum panels to provide comfortable secure penning at night. Pens with electrified fencing and a solar generator can be used during the day.

Sheep are moved most effectively from the pens to the grazing area, or between grazing areas, when assisted by a herding dog. Border collies are popular and effective sheep dogs and are relatively cheap and easy to obtain. Properly trained dogs should, as a minimum, be able respond to commands to circle the sheep in both directions, lie down and walk up to the sheep to force them forward. Dogs that bite, harass or in any way worry or excite the sheep should not be used in woodlot grazing management.

If predators such as coyotes or wolves are present in the woodlot you may want to consider a guardian dog such as a Maremma. These dogs can be trained to work alone or in a group. The guardian dog precedes the flock and in the case of a group of dogs, others will guard the flanks. The properly trained guardian dog will only rest when the sheep are grazing or are safely penned at night.

Predators may be detected and discouraged from a grazing area by following a few preventive measures. It is important every day to walk the area to be grazed to check for predator signs before releasing the sheep. Allowing the sheep dogs or guardian dogs to mark scent posts around the grazing area to claim the territory will often discourage predators.

Keeping predators away may be achieved by having no food association in the grazing area. Removing all garbage, spilled dog food and animal carcases (no burying or burning) will remove the scent of food for the predators. Sick or lame sheep stimulate predator attack and should also be removed.

Sheep are ruminants, which means they have two stomachs. The first stomach receives the food and uses bacteria to partially digest the forage. After the first stomach is full the sheep rechews the contents of the first stomach (chewing of the cud) and swallows it into the second. The sheep feeding cycle goes through two specific events, first the sheep feeds for the woodlot manager (grazing), second, the sheep feeds for itself (cud chewing).

Maximum productivity can be obtained when the sheep have two specific grazing periods separated by two rest periods where the animals have access to water, salt, vitamins and minerals and where the cud can be chewed. This feeding schedule is superior to letting the animals out to graze all day because by allowing specific rumination times, the sheep's stomach becomes synchronized and consumption is increased. Maximizing the flock appetite stimulates greater grazing impact on a wider range of vegetation during the grazing period.

If sheep are only to be moved from one field to the next, it is not imperative to provide special conditioning. If the sheep, however, are to be trucked a great distance to the reforestation site, it then is necessary to acclimatize them to their new environment. The stress caused by change in diet, trucking, mixing of flocks and disorientation at the new site may compromise the condition of the flock and decrease grazing performance.

Removal of competing vegetation allows unrestricted growth of newly planted (left) and established seedlings (right).

Transportation and orientation stress can be minimized by gradually tempering the sheep to their new environment, for example, with shorter, more frequent grazing periods supplemented by low- protein hay to gradually increase rumen capacity for the native forage. Initially, a palatable high carbohydrate supplement such as a molasses block should be provided to increase energy in the short-term until the sheep have settled into a specific grazing schedule.

For maximum vegetation control, a program should be set up to graze the sheep over the same area on a number of occasions, or "passes." The number of passes made over an area depends on desired level of impact, the vegetation complex, precipitation and the distance between grazing sites.

A typical goal for the woodlot manager is 75% removal of undesirable vegetation with 85% of the crop trees at least 50% higher than the competing vegetation. In dry conditions this type of impact target can be obtained by using the two-pass system.

The two-pass system allows the area to be grazed for one or more days (first pass) then allowing two days before the area is again grazed for the second pass. Leaving the trampled and lodging vegetation to wilt for two days reduces its palatability and because only vibrant plants need to be grazed, the second pass is quicker and more effective. In areas with a moist climate, it may be necessary to implement the two-pass operation more than once during the summer.

The number of sheep required for a given reforestation program depends on a number of factors, including local climate, competitive species present, topography, breed of sheep and number of passes. The rule of thumb for a minimum stocking rate is four sheep per hectare (16 sheep per 10 acres). In very small reforestation programs, flocks can be borrowed from a neighbour for a few days and then returned.

If large flocks are confined overnight in small pens, it may be worthwhile to recover the sheep manure while the sheep are out grazing. The sheep manure is nutrient rich and is an excellent fertilizer when added to the seedling potting soil.

Tree Trivia

The easiest way to determine the sheep's age is by its teeth. A lamb has eight small incisor teeth until it reaches one year of age. Each year one pair of lamb teeth is replaced by two permanent teeth which are considerably larger. By the time the sheep is four years old, all of the lamb's teeth have been replaced and age is determined by tooth wear. Usually the sheep wears out its teeth at about seven or eight years old and from then on the sheep has to gum it!

Horse Logging

Horses used in woodlot management can be a practical and economical alternative to operating with machinery. Horses have been used for moving logs from the forest since Canada's very early days. Horses are ideally suited for skidding logs from the falling to the landing areas or just hauling around equipment in rural and urban woodlots. Horses are most beneficial where space for logging roads is at a premium, in environmentally sensitive areas, and where tourism is an important aspect of the business.

Horse logging is most cost effective for commercial thinning and selective harvesting operations. The narrow width required by horses to skid logs can reduce or eliminate the cost of building trails or roads.

The horse can also have a much lower environmental impact than machines in certain conditions. The use of horses minimizes soil degradation and erosion potential, making horse logging ideal for removing trees from environmentally sensitive areas such as riparian zones.

Farming with horses is a tourist attraction associated with other types of home-based businesses such as market gardens and stables that offer hay rides. Because the horses have year-round "fixed" costs such as feeding and watering, having them log in the off-season can make economic sense and keep the horse's muscle tone year-round.

Although the use of horses in woodlots has many advantages, it is important to consider all the implications of horse ownership before you venture into this type of management. First, logging with horses is both physically demanding and time consuming. Secondly, horse logging requires individuals with prior experience in both horses and logging to avoid injuring or killing either the horse or logger. People without prior instruction or with little back-

In the early days of logging, horses were the only method of hauling logs from the bush. (Courtesy Lake of the Woods Museum.)

Logging with horses has not changed too much from 100 years ago. It can stll be tough, cold work. (Courtesy Lake of the Woods Museum.)

ground working with horses should either apprentice with an experienced horse logger or take formal training at an accredited institution.

Horses, unlike vehicles, cannot be parked when they are not in use. They require a basic infrastructure including a barn or shed, fenced exercise area and a continuous supply of quality hay and water. As well, horses require considerable time and effort from the owner for training and stall cleaning for example.

Most people associate large draft horses, which weigh 900 kg (2000 lbs) or more, with log hauling. In fact, it is usually an advantage to use smaller breeds that weigh in at around 450 kg (1000 lbs).

The most popular large breeds for horse logging are the Belgian and Percheron because they lack the long hair or "feathers" on their legs that tend to accumulate mud, snow, sticks and other debris. Purchasing a horse previously trained to haul logs is usually the best bet for the budding teamster, although just about any well-conditioned draft horse between three and ten years old can be successfully trained.

If the horse is to be used for logging purposes for only a small fraction of the time, then smaller "riding" breeds such as Arabians, Morgans, Irish Hunters, etc., may be used. The advantage of using smaller horses is that when not in use they use only half the feed of the larger horses. Smaller horses can also be used for recreational riding when not hauling logs.

Horse logging requires some specialized equipment above and beyond the halter, bridle and bit that are common in Canadian barns. Special logging harness consisting of traces, back and belly band, backstrap and cupper, tugs and breeching are required. The rigging for log hookup is also necessary.

After the horse and harnesses have been purchased, a three- to six-week "breaking-in" period is necessary for both the horse and the teamster before the horse-logging operation can be expected to work a full eight-hour day. The breaking-in period is essential for three reasons: to physically condition the horse, to harden the animal's hide against the friction of the harnesses to prevent blistering, and for horse and teamster to get acquainted, so that each will know each other's limitations and what is expected from the other in strength, commands, etc.

These Percherons can haul larger logs but have limited recreational use once the logging season is over.

Care of the horse does not end with the breaking-in period. In order for the horse to perform to its maximum, proper nutrition is required and should include the correct amounts of grain, hay, water and minerals. Do not always conclude that sufficient quantity and quality of fodder is available on the forest floor. When in doubt haul in additional supplies. A rule of thumb for feeding is 1.1 kg (2.5 pounds) of grain and 1.25 kg (3 pounds) of hay for each 100 kg (220 pounds) of horse per working day.

Of all the daily tasks such as brushing the animal down and treating minor wounds and fly bites, probably the most important area of care is the hoof. Even if the horse is not being worked on rocky or broken ground, it is important to examine each hoof daily. The hoof should be cleaned as necessary and broken pieces rasped as they occur. Although horse shoes are not necessary for horse logging, they should be considered if hoof deterioration becomes apparent.

The logging operation should be well planned before the arrival of horses. The trails should be selected on the basis of minimum grade and side slope. Stumps should be cut close to the ground to avoid damaging the lower legs, hence great care must be given when using horses to skid logs from beaver-logged areas.

If the logging operation is to be carried out during the big game hunting season, it is prudent to take additional precautions against accidental shooting of the horses. Access to the logging area should be well marked with highly visible, clearly readable warning signs (SEE NEXT PAGE). Fluorescent flagging attached to the harnesses and rigging will also provide an additional safeguard.

It is important to recognize that horses do not have the power of many machines used in traditional logging operations. Brute force is not always an alternative in horse logging; therefore the teamster's savvy

These Morgans are suitable to haul logs, but can be used for pleasure riding and only require half the food for maintenance in the off-season.

and horse manoeuvrability must be used instead. A skilled teamster understands the limitations of the horse and adjusts the logging practices accordingly. These adjustments include log size, trail slope and working conditions.

Once teamster and horse become accustomed to the working conditions and to each other, high volumes of production are attainable. A horse can be so well trained that it becomes unnecessary to walk with the horse from the logging to the landing areas. In this case the logger attaches the log to the horse's harness and the horse follows the trail to the landing area by itself where another person detaches the log, and turns the horse around to pick up another load.

Horses used in forest management are likely to have their greatest potential in urban areas where all the positive aspects of horses can be used. Urban forests often include narrow trails that are maintained primarily for recreational purposes such as walking or biking. Like well-managed rural woodlots its necessary to remove trees that are diseased, have fallen across the trail or pose a safety hazard to trail users. Horses are ideally suited for this type of forest maintenance.

As well many cities already have mounted police officers to patrol city parks, partly because the horse is versatile enough to be effective in this environment and partly because the mounted police are

Since horses are similar in appearance to many big game animals, it is important to warn hunters that horse logging is in progress.

a great tourist attraction. Horses used for maintaining urban forested areas also exhibit this versatility and tourist appeal.

Many urban forests grow in parks situated on riparian areas (SEE PAGE 118) and other environmentally sensitive areas. Because horses have minimal environmental impact they demonstrate another distinct advantage in urban forest management over traditional forestry practices.

The City of Winnipeg is one of the first Canadian cities to recognize the economic and environmental benefits of using horses in urban woodlot management. As well, the Alberta Pacific pulp mill has initiated a pilot project to determine the feasibility of horse logging on otherwise inaccessible stands. Horse logging may be in high demand as more timber companies and urban areas acknowledge the many benefits of using horses in their forest management.

Tree Trivia

Horses are gentler, consume less water and are under less stress when trees tall enough to provide shade are within the pasture. The horses are cooler and the face fly problem is reduced when horses are in the shade.

Planting For Wildlife

For many woodlot owners, the primary goal for the family forest is enhancement of wildlife habitat. Such aspirations are admirable, but it is also important to realize that management for wildlife is very much species specific, that is, no one management strategy will attract all animals. An animal will be attracted to particular woodlot if it contains the five essential elements of its habitat: water, food, shelter,access and space.

Water is very important in attracting wildlife as anyone with a birdbath in the backyard will attest. Woodlot water is important as a drinking supply and provides nesting areas for ducks, an abundant supply of plants and animals for a food source, and a cool place for a mudbath. The list is virtually endless. For fish and aquatic mammals such as beaver, muskrat and otter, water is absolutely essential as part of their habitat.

Often a forest area is fortunate to have a naturally existing body of water such as creek or a pond. Special management techniques must implemented to maintain water quality. Improper harvesting or roadbuilding practices may result in massive soil erosion, which can fill ponds and sloughs and choke flowing streams. Unrestricted cattle grazing may also cause stream and pond degradation by collapsing banks, stirring up mud and contaminating the water with fecal material.

Water body and associated riparian zones.

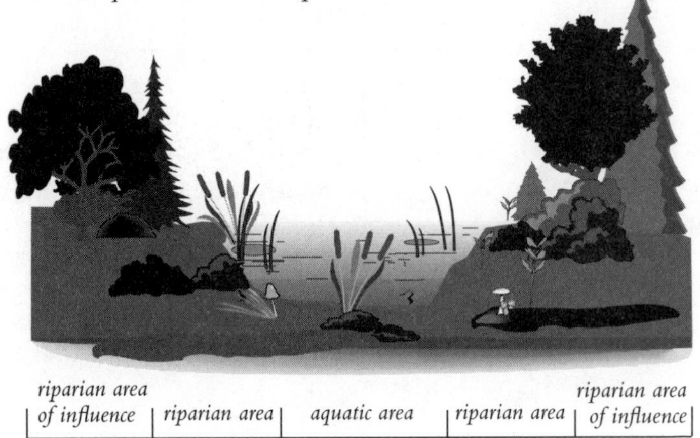

In the case of flowing water, maintaining a buffer zone of trees, shrubs and other vegetation along the stream margins (riparian zone) will do much to mitigate the negative effects of other activities in the forest. The plant's roots holds the soil and stabilize the banks against the erosive force of moving water thus maintaining areas of deep water (pools) for fish and other aquatic animals. The bordering vegetation also provides shade, which helps maintain water temperatures below critical levels for fish survival.

Ponds in a woodlot are managed differently than flowing water. Unless the body of water is very large and wave action is anticipated to cause erosional problems or the banks leading to the pond are very steep, it is not as critical to maintain tree growth around the entire shoreline. Small cleared areas around the pond may, in fact, attract wildlife as they provide the animals with a clear view of approaching predators and allow greater wind circulation to assist in oxygenating the water.

If water is not present, artificial sources can be built easily and inexpensively. For the urban woodlot and other small stands of timber, small pre-formed plastic or cement drinking troughs are usually adequate. People who would like a large pond for attracting ducks, stocking fish or providing habitat for a beaver-logging operation can obtain the book *Pond Raising Rainbow Trout*.

Food is another critical component in attracting wildlife to a woodlot. Remember that each species has a slightly different food source, and any habitat change will have a positive effect on some animals and a negative effect on others.

A source of water such as a creek or lake is very important in attracting wildlife. This pond was constructed by building a small earth dam with a backhoe.

Game animals such as deer and elk thrive in fragmented forests with alternating small clearcuts and mature stands. The full sunlight stimulates the grow-th of groundcover for grazing and saplings for browsing.

The fragmented forest also provides "edge," the strip of land along the forest edge where animals can graze yet still have cover close by in the event of predators. Farm shelterbelts, from a wildlife point of view, are continuous strips of edge without the rest of the forest. Edge is the habitat of choice for many species of small game such as rabbits and grouse.

Predators such as owls and hawks are often found on the edge because it offers a perch with a large field of vision. Carnivores such as foxes and coyotes also use the forest edges as their primary hunting grounds because of the abundance of wildlife.

Continuous overmature climax forests produce very little food for large grazing animals. In many national and provincial parks, for example, fire control and logging prohibitions have resulted large climax forests. The large ungulates, unable to find food elsewhere, come to graze in the open areas along the major highways.

Climax conifer forests produce an abundance of cones that squirrels eat. The squirrels in turn are used as a food source by predators such as pine martins, fishers, weasels and goshawks. Dead standing trees (snags) in the climax forest also provide a source of food for woodpeckers.

Snags also provide shelter for many types of nesting and denning animals using the holes and burrows created by the squirrels and woodpeckers.

Sheltering wildlife from the elements and from other animals is very a important aspect of the forest ecology and varies according to the animal's needs. Deer, for example, tend to select areas that offer thermal protection in the winter such as conifer groves with closed cano-

Many deer and other animals feed along grassy roadside ditches because road construction has created an artificial forest edge.

pies to provide a wind shield and a refuge from deep snow. Turtles and frogs, on the other hand, hibernate during the winter and require partially submerged logs for basking on and hiding under in the summer.

Shelters such as birdhouses or artificial ponds can be cheaply and easily constructed to enhance existing woodlot features. Drilling out cavities in snags or providing nesting boxes will encourage bird species that might otherwise bypass your woodlot. A large round straw bale placed on end in a pond or slough will encourage the nesting of Canada geese.

Fish require shaded "pools" where stream velocity is decreased and depth is increased. These pools occur naturally behind fallen logs and downstream of waterfalls, but they can also be created by adding large rocks or excavating the stream bottom.

Access into the woodlot is important for animals and humans, which both tend to avoid areas with improper access because of the amount of energy required to move about. Often in a multiple-use woodlot, animal access is a byproduct of other forestry operations such as logging or fire control.

Forests with poor access usually fall into two categories: excess deadfall and interlocking branches. Excess deadfall comes from forests that are overmature, diseased or support a population of beaver. Forests with interlocking branches are either too densely restocked or restocked with primarily brushy vegetation.

Cutting trails is usually the most economical way to maintain access for animals. Trails benefit the wildlife and allow a greater degree of forest management and fire control. Construction of trails is outlined in Chapter 4.

Dead standing trees (snags) provide shelter for a wide variety of birds and animals. This hole, originally formed by a woodpecker, is now the nest of a wood duck.

This blue heron rookery is situated in the uppermost tree branches. Sites with a good view and a close source of water are where herons prefer to build their nests.

Creating access by trail construction can have spectacular results. A virtual animal highway can be created when the trail connects adjacent feeding and watering sites or joins two segments of the animal's territory.

The last but certainly not the least aspect of habitat is space. Wildlife are territorial by nature and require a certain amount of space in order to co-exist with others within and without its particular species. The space required by each animal is again determined by the individual itself. A pair of squirrels, for example, may be completely happy in a small grove spruce trees, whereas larger animals such as deer, elk and caribou require a larger area.

Obviously the larger the size of your woodlot, the larger variety of animals that are likely to live there. But even a small woodlot can become part of a particular animal's large territory by providing the other four elements of habitat: water, food, shelter and access.

The smaller the woodlot the smaller the population of individuals it can support. Animals that have become overpopulated in a given area will resort to emigration, fighting and even cannibalism to restore the population to self-supporting levels. For example, two or three beavers may be transplanted into a new area to initiate a beaver-logging program, but transplanting two or three dozen beavers

This artificial nesting box has attracted this family of horned owls.

Artificial pools can be constructed in flowing streams by periodically laying large tree trunks at right angles to stream flow.

will create such an intense competition for space that many of them will emigrate or die.

To attract a particular species of animal to your woodlot, it is necessary to provide habitat suited to that animal. An animal's habitat consists of the five elements described above, and if any one of these elements is missing, that animal will only use the woodlot on a discontinuous basis.

Tree Trivia

The most common animals you will likely "see" in your woodlot are birds. Although some birds make regular appearances, many birds will only be known for their song. Birds "singing" is a method of marking territory. If another bird enters that territory it will get an earful from the resident. Birds can be tricked into coming close to a human by a process called "pishing." By softly repeating the word "pish" and holding the sh (pishshshsh... pishshshsh... pishshshsh), the songbirds, thinking an intruder is present, will fly closer to investigate. Most pishers are early risers because the greatest songbird activity is right around sunrise.

Smaller animals such as this red squirrel require less space per animal and therefore are common in woodlots.

CHAPTER NINE

A REAL LIFE WOODLOT HARVEST EXAMPLE

The following photos are the sequence of events that resulted in the harvest of .6 hectare (1.5 acres) of trees in my 60 hectare (125 acre) woodlot. The area logged was clearcut of all trees with a dbh of greater than 15 cm (6 inches). The clearcut area comprised a strip 15 m (48 feet) wide and 400 m (1200 feet) long.

April 1994. *A mature stand of aspen on the west side of my field had begun to die off. The dead trees would be blown onto the field by strong winds, hampering agricultural field operations and necessitating removal after every storm.*

October 1994. *View of the field before harvest.*

Trees after falling.

November 1994.
Trees were skidded.

View of log decks.

December 1994. *Logs were sawn into lumber.*

View of sawing operation.

March 1995.
View of sawmill residue.

April 1995. *Logs not sawn were sold to the Alberta Pacific pulp mill.*

May 1995.
Sawdust was loaded and used for mulch on nursery trees.

The dead standing trees were cut into firewood and loaded onto the truck with a modified bale stacker (left) and stored underneath a pole barn made from last year's timber harvest (above).

June 1995.
The lumber, dry piled in March, is ready for use (SEE PAGE 95).

July 1995.
The field has been returned to conventional agrarian pursuits. Note regeneration on cut area.

Tree Trivia

The official accounting of the above harvest operation is as follows:

REVENUE		
Commodity	Amount	Value
Lumber	*6500 board feet*	*$1950.00*
Firewood	*12 cords*	*$ 600.00*
Pulpwood	*24.92 tonnes*	*$ 274.12*
TOTAL REVENUE		**$2824.12**

EXPENSES		
Fuel	*250 litres*	*$ 100.00*
Chainsaw oil	*10 litres*	*$ 12.00*
Sawmill rental	*2 days*	*$ 450.00*
TOTAL EXPENSES		**$ 562.00**

Woodlot Timetable

January

- Too cold and dark to do anything outdoors
- Ideal for indoor woodworking or using firewood

February

- Saw decked timber
- Haul any remaining wood from field to processing facility

March

- Tap maple trees
- Dry pile remaining sawn lumber

April

- Volume and density sample plots
- Thinning treatment for fenceposts and rails

May

- Tree planting
- Vegetation control

June

- Trail maintenance
- Mushroom harvest

July

- Enjoy the woodlot with your family
- Wildflower harvest

August

- Berry picking
- Trail construction

September

- Cone and nut collection
- Pond or dugout construction and seeding

October

- Timber harvest
- Skidding and decking

November

- Collecting shoots for vegetative propagation
- Cutting and stacking firewood

December

- Conifer thinning for sale as Christmas trees
- Pruning, boughs used for Christmas wreaths

GLOSSARY

Actual size: The size of a board after it has been dried, planed and edged; the metric measurement is the actual size converted to metric units. See also nominal size.

Agroforestry: The combined practices of agriculture and silviculture for mutual benefit.

Apical bud: The top part of the crown of a tree responsible for upward growth.

Artificial regeneration: Planting trees.

Asexual reproduction: Results in production of saplings from the roots (suckering). Also called "vegetative reproduction."

Auxin: A hormone that occurs in the tree's crown and trunk; inhibits suckering. See also Cytokinin.

Beaver fever: See Giardiasis.

Board foot: Standard measurement for lumber equivalent to a piece of wood 1 foot square and 1 inch thick; abbreviated as FBM.

Cambium: The actively growing cell layer of a tree's vascular system.

Capybara: Largest rodents in the world; from South America.

Chlorophyll: A substance in leaves that provides the green colour to green leaves; absorbs other wavelengths of light that are most effective in photosynthesis.

Clearcut: Harvest method that simulates a naturally occurring catastrophic disaster.

Clear wood: Knot-free wood.

Climax forest: See Old-growth forest.

Commercial thinning: Thinning that produces trees with sufficient size to be commercially valuable.

Conifer: Tree characterized by needle-like leaves and seed-carrying cones. Also called "evergreens."

Conifer release: See Release.

Cord: A pile of stacked wood containing 128 cubic feet of wood, bark and air (4 feet wide by 4 feet high by 8 feet long).

Crown: The upper part of the tree containing the branches and leaves.

Cycad: Palm-like plant present in New Brunswick and Nova Scotia 320 million years before present that contributed to those provinces' coal beds.

Cytokinin: A hormone that occurs in the tree's roots; promotes suckering. See also Auxin.

Daughter trees: Saplings produced from the roots by asexual reproduction; genetically identical to the "mother" tree.

Dbh: See Diameter at breast height.

Deciduous: Broad-leafed tree characterized by summer production of leaves followed by leaf loss before winter, and a seed protected by a capsule.

Decked timber: Trees that have been fallen, delimbed and piled in a landing, field or other truck-accessible area.

DED: See Dutch elm disease.

Diameter at breast height: The tree diameter that occurs about 1.5 m off the ground. Also called "dbh."

Dry pile: A method of stacking lumber to allow the sun and wind to remove moisture from the lumber.

Duff: Large amounts of organic matter on the forest floor.

Dutch elm disease (DED): A devastating insect infestation of elm trees.

Edge: The edge of the woodlot where wildlife often forage or graze on succulent new growth with protective cover close by.

Emergency Response Strategy (ERS): A set of steps to follow in the event of an accident or natural disaster.

Fatal 15: The hazardous 15 seconds between the time the tree first begins to fall until it is safely on the ground and all debris has fallen.

First lift: Pruning branches so that the trunk is clear for the lower 3 m.

Forest Products Hauling Record (FPHR): Necessary if transporting logs on municipal or provincial roads and necessary to ensure that loggers are not illegally harvesting private or crown land.

Forest succession: The natural process of changes that occur in a forest's species.

Forestry diameter tape: A measuring tape calibrated to show units of diameter of a tree trunk.

Giardiasis: A disease that causes extensive diarrhea and abdominal cramps brought about by water contamination caused by beaver overpopulation. Also called "beaver fever."

Ginkgo: Trees that were present in New Brunswick and Nova Scotia 320 million years before present that contributed to those provinces' coal beds.

Greywater: Nutrient-rich water coming from clothes washers, showers and baths.

Growth rings: Annual layers of light and dark sapwood that help determine the age of a tree.

Hardwood: Timber derived from deciduous trees.

Heartwood: An inner layer of the tree composed of sapwood that has lost the ability to conduct fluids and dissolved nutrients.

High grade cut: The practice of harvesting only the biggest and best trees. Also called "high grading."

Homogenous Unit: A small area within the woodlot that has similar tree density, age and species for the purposes of timber volume and value estimates. Also called "Huey" and HU."

Inner bark: See Phloem.

Intarsia: A wood art involving making a picture by cutting out parts of an original design from different colours of wood, polishing them and combining them into an intricate inlay that fits together like the pieces of a puzzle.

Juvenile spacing: Thinning to prevent overcrowding of very young stems. Also called "precommercial spacing."

Leader whip: Results when continual contact of the apical bud causes competition from lateral branches and possibly the formation of multiple leaders.

Limiting factor: A shortage of one of the three essentials to photosynthesis (light, water and carbon dioxide).

Monoculture: Forest with only one species of tree.

Mother tree: The genetic source of clones produced through asexual reproduction.

Natural regeneration: Nature handles the reforestation process.

Nominal size: The dimensions of a rough green board before planing and edging, the unit of measurement used by most retail lumber yards.

Nurse crop: Species of tree that dies off after light becomes restricted by a higher canopy of trees; nutrients from the nurse crop are returned to the roots, where they can be conserved for many years until light conditions allow the plant to sucker successfully.

Old-growth forest: A forest that has escaped disturbance for a long time; the climax forest in a forest succession dominated by large diameter spruce and balsam fir. Also called "climax forest."

Pace: Two normal steps, usually the same length as your height.

Phloem: The layer of a tree inside the outer bark; conducts food from the crown to the other parts of the tree. Also called "inner bark."

Photosynthesis: Process by which plants convert carbon dioxide in the air into sugars and related substances.

Pioneer tree: Shade intolerant species that occurs in the first stage of forest succession; e.g., poplar.

Plugs: Plastic or styrofoam trays filled with high nutrient soil used for growing seedlings.

Potipuki: A device used for planting plug seedlings without having to bend over.

Precommercial thinning: See Juvenile spacing.

Pseudo-natural regeneration: Preparing the site for reforestation.

Regeneration: See Artificial regeneration, Natural regeneration, Pseudo-natural regeneration.

Release: A process involving the removal of competition from other species, allowing rapid growth indicated by widely spaced growth rings. Also called "conifer release."

Riparian zone: Buffer zone of trees, shrubs and other vetetation along stream margins.

Rollover protection structure (ROPS): Typical of minimum tractor safety features for woodlot work.

Roots: Like underground branches of a tree, responsible for anchoring, nutrient uptake, storage, etc., and sometimes reproduction.

Roundwood: The solid volume of wood that occurs in a log or group of logs; roundwood does not include bark or airspace and is measured in cubic metres.

Ruminants: Animals such as sheep with two stomachs.

Salvage cut: A harvesting technique involving removal of dead or dying trees that have succumbed to wind, fire, disease or other maladies.

Sapwood: See Xylem.

Seed tree cut: Continuous commercial thinning of a stand where only a few of the largest, most desirable trees are left to produce seed for the next generation of trees.

Selection cut: The individual removal of certain designated trees in a forest.

Shade intolerant: Description of a species that does not grow well in shade; e.g., pioneer species such as poplar.

Shade tolerant: Description of a species that grows well in shade; e.g., old-growth species such as white spruce and balsam fir.

Shelterbelt: A dense windbreak of trees that helps reduce heat loss in homes, reduce soil erosion, trap snow, increase returns in livestock and crop operations, and increase wildlife populations.

Shelterwood cut: The practice of leaving certain trees standing for the protection of subsequent generations of trees.

Silviculture: Latin for "forest cultivation."

Silvipasture: The practice of grazing animals in a forested area.

Slab: Board with one sawn side.

Slab board: Board with two sawn sides.

Slash: Debris left on the woodlot floor after harvest.

Snag: Dead standing tree, often used by wildlife.

Softwood: Timber derived from conifers.

Stacked wood: Includes wood, bark and airspace and is measured by the cord.

Stand: A community of trees that exhibit common characteristics such as species, age, height or diameter.

Standing timber contract (STC): An agreement between a woodlot owner and a buyer who becomes responsible for aspects of the harvesting operation.

Stereoscope: A device used to view air photographs that provide overlapping images showing relief (three dimensions): e.g., tree height, valley contours.

Stomata: Sausage-shaped openings in a leaf through which carbon dioxide enters.

Suckering: See Asexual reproduction.

Thinning cut: Performed in stands where overcrowding prohibits obtaining maximum value of the stand; two types: precommercial thinning and commercial thinning.

Timber cruising: The enjoyable ramble in the woodlot to do timber volume estimates; can be conducted by the whole family or just you and the family dog.

Trunk: The body of the tree, providing strength to support the crown and containing the pipelines for food and fluid movement within the tree.

Vegetative reproduction: See Asexual reproduction.

Wilding: Natural seedling.

Workers' Compensation Board (WCB): Has developed safety requirements that must be met by woodlot workers.

Xylem: Layer inside the cambium that carries water and minerals from the roots to the crown. Also called "sapwood."

ABOUT THE AUTHOR

Bruno Wiskel, unable to hold down a regular job, has reverted back to nature on his farm near Colinton, Alberta, where he raises fruit, vegetables, rainbow trout and other traditional crops in the summer. In the winter he is a writer and operates an environmental consulting company called Evergreen Environmental Ltd.

Bruno shares his farm with his two faithful dogs Sasha and Pork Chop, who head up the "Small Animal Chasing" department of Evergreen Environmental and hope to be promoted to the "Small Animal Catching" department.

Bruno's first book *Pond Raising Rainbow Trout* was a blockbuster hit with small-scale aquaculturalists across North America.

Sales of *Woodlot Management*, Bruno's second book, are expected to surpass even *Pond Raising Rainbow Trout*.

The author with his dogs Sasha (left) and Pork Chop (right) on a pile of freshly sawn lumber.